TRANSFORMATION OF HERBICIDAL TECHNOLOGY

Chemical Based to Ecological Concern

Dr. Rudraraju A. Raju
M.Sc (Ag), Ph.D, FFSRDA
Acharya N.G. Ranga Agricultural University
A.R.S., Maruteru-534 122, India

2018
Daya Publishing House®
A Division of
Astral International Pvt. Ltd.
New Delhi - 110 002

First Published, 2001
Reprinted, 2018

ISBN: 978-81-7035-236-5 (Hardbound)

Publisher's note:

Every possible effort has been made to ensure that the information contained in this book is accurate at the time of going to press, and the publisher and author cannot accept responsibility for any errors or omissions, however caused. No responsibility for loss or damage occasioned to any person acting, or refraining from action, as a result of the material in this publication can be accepted by the editor, the publisher or the author. The Publisher is not associated with any product or vendor mentioned in the book. The contents of this work are intended to further general scientific research, understanding and discussion only. Readers should consult with a specialist where appropriate.

Every effort has been made to trace the owners of copyright material used in this book, if any. The author and the publisher will be grateful for any omission brought to their notice for acknowledgement in the future editions of the book.

Published by : **Daya Publishing House®**
A Division of
Astral International Pvt. Ltd.
– ISO 9001:2015 Certified Company –
4736/23, Ansari Road, Darya Ganj
New Delhi-110 002
Ph. 011-43549197, 23278134
E-mail: info@astralint.com
Website: www.astralint.com

Dedication :

To

Sri N. Chandrababu Naidu

The Man who is a product of modern A.P., intellectual disciplinarian, redeemer of rededundant

Swarnadhra Pradesh

With a spirit of renaissance through techno-based, social and economic revolution.

What is on earth,
Nay, under heaven, continues at a stay ?
Ebbs not the sea, when it hath overflowed.
Follows not darkness when the day is gone ?
And see we not sometimes the eye of heaven
Dimmed with overflowing clouds ?
There's and that work of careful nature, of cunning art,
How strong, how beauteous,
or how rich it be,
But falls in time to ruin.

—*William Shakespeare*

Don't believe anything
Ask yourself
Why ? What ? Where ? and How ?
Truth will be visible
The right answer will spring
from the churned mind.

—*Goutama Buddha*

Foreword

Weeds are undesirable plants which interfere with the utilisation of land and water resources and thus adversely influence human welfare. They compete with crop plants for nutrients, soil moisture, space and solar radiation and adversely affect photosynthetic activity and productivity of crop communities. Depending on the degree of competition, weeds reduce the crop yield by 10-90 per cent. Thousands of hectares of cultivable area in the crop growing regions of India have been abandoned due to severe infestation of perennial grasses. Contamination of noxious weed seeds greatly reduced the value of crops and renders them unsaleable. Weeds also reduce human efficiency through physical discomfort caused by allergy and poisoning.

Weed management is as old as agriculture itself. It continues to be of immense interest and great concern to the researcher and farmer alike in the continuous struggle for maximising crop production to meet the growing needs of burgeoning population. Herbicidal technology has turned relevant for highly efficient and time saving weed control.

There has been much controversy in the past two decades concerning the effect of herbicides on the environment. Herbicides kill the target plants, but leave residues which may persist in the environment either in the parental form or metabolites to affect the organising through their toxic effects. Some herbicides are acutely toxic to man, animals and wildlife. continuous use of any method of weed control may raise many questions about the possible undesirable side effects. The most dependable results in weed control are obtained by using as many of the various methods of weed control as possible in any given situation. There is a growing evidence in support of "second generation" problems by means of shift in weed flora due to continuous use of closely related herbicides.

The need for a comprehensive and updated treatment of the subject of weed science other than chemical weed control. I am happy that Dr. R. A. Raju has painstakingly written a reference-cum-textbook to meet the present needs. The book has some valuable chapters like weed control in cropping systems, integrated weed management, herbicidal persistence and human health and allelopathy.

The present work will have a great impact in the world of weed science and stimulate the minds of researchers, teachers and students of agriculture and allied fields. I congratulate the author on the timeliness of writing this useful document which is long overdue.

M.V. RAO

Ex-Special Secretary, DARE, Govt. of India

Ex-Special Director-General, ICAR, New Delhi

Ex-Vice Chancellor, A.P. Agricultural University

Chairman, BPC, APNL Biotechnology Programme

President, Andhra Pradesh Academy of Sciences

Hyderabad

Preamble

Weeds have been a persistent problem for farmers ever since agriculture began. They are not only major constraints in crop production but also create problems in non-cropped lands world-wide. In traditional crop production, weeds are managed by hand weeding. But today's fast industrialization in many developing countries offers the labour force an opportunity to earn more money outside the agricultural sector, and labour is in short supply for hand weeding.

Shortages of labour is putting pressure on farmers to change their cultural technology. With the result, weed control was shifted from hand weeding to chemical application. Use of herbicides is becoming new weed management strategy in several crops. Herbicide consumption is rapidly increased in all Asian nations during the past decade. Herbicides are easy to use, reliable and quick method of controlling weeds. Heavy and continuous use of chemicals may pose serious problems of pollution of the environment. Besides contamination of soil and water resources, development of herbicide resistant weeds and human health problems are increasing year to year due to chemical weeding. In view of the afforesaid problems it is necessary to diversity weed management options.

Farming is undergoing rapid changes in tropics and subtropics to meet the needs of burgeoning population. Scarcity and high wage rates of labour in rural world resulted in searching of other ways of weed control. Research must find ways for the sustainable management of weeds as the new trends moving at a rapid pace. Relying on chemicals may have many of the consequences of earlier experiences with other pests. The only holistic approach to control weeds without any ecological disturbances is integrated weed management which is ecofriendly and economically viable.

The primary reason for writing this book has been to emphasis the principles and basic aspects of weed control other than chemical use. It is based on my personal acquintance of weed management in Indian agriculture. The weed management strategies planned in developed and effluent countries cannot be suited to high populous country such as India. The developed world have a inborn tendency to turn the limited resources into luxurious commodities which results in global pollution. I have been influenced by 20th century legends such as Mahatma Gandhi and Swami Vivekananda who preached self-restraint in human desires. The conservation of natural resources is merely not sentimental but the rediscovery of a truth well-known to our ancient sages. They teaches us that all forms of life—plants, animals and humans—are closely linked, so that, disturbance in one gives rise to imbalance the other.

This book could not have been completed but for the enthusiastic support of numerous colleagues and friends, both inside and outside India. My heart-felt thanks goes to Dr. S. Alstrom, Swedish University of Agricultural Sciences, Uppasala, Sweden; Dr. Keith Moody, Hawaii; Dr. P.C. Bhowmik, University of Massachusetts, Ammherst, U.S.A.; Dr. V.M. Bhan, Director, National Research Centre for Weed Science, Jabalpur for their support and competent assistance.

My understanding of weed science and insight on the Indian scene came from my interaction with several national scientists and participation with progressive farmers. My heart-felt thanks to my spouse, Lakshmi Rudraraju and other family members as they released me from domestic responsibilities during my work on the script so that I could get this book out.

I hope that this book is useful as a text-book to the students of agricultural sciences at under-graduate and post-graduate level, as well as reference book to teachers, researchers, extension officers of government and non-government organizations and herbicide industry. I humbly conclude with the following quatation of *Mahaveer Jain*, the great preacher of compassion :

One cannot enjoy the natural world with greedy mind,
He should have an insight about laws of nature,
The man cannot be called as civilized human
unless he looks upon all creatures of
God with the eyes of a friend.

Rudraraju A. Raju

Contents

	Foreword	xi
	Preamble	xiii
1.	Introduction	1
2.	Genesis of Soil and Xerarch Succession	8
3.	Soil—A Prominent Abiotic Adaphic Factor	14
4.	Change of Weed Flora Due to Agro Ecological environment	27
5.	Shifts in Weed Flora	38
6.	Weed Management in Cropping Systems	46
7.	Impact of Herbicide use on environment	51
8.	Herbicide Persistence and Human Health	60
9.	Social and Ecological Implications	67
10.	Integrated Weed Management	75
11.	Allelophthy—a biological warbare	99
12.	Barnyard Grass—A Mimetic in Rice Fields	109
13.	Noxious Wild rices	121
14.	Menace of Red Rice	127
15.	Parasitic Weeds	132
16.	Soil Solarization	149
17.	Management of Obnoxious Weeds	158
18.	Flooding as Weed Suppressing Tool	173
	Appendicies	188

Appendix 1. — Noxious weeds-in brief

Appendix 2. — Cultural weeding with animal operated ploughs

Appendix 3. — Scientific names of economic crops

Appendix 4. — Problematic weed seed present in crop seeds

Appendix 5. — Some Exotic weeds introduced into India from Various contries
Appendix 6. — Weeds Adentified as Quarantine pests
Appendix 7. — Prominent weeds of India
Appendix 8. — Ability of Tolerance of Wild rices
Appendix 9. — Converson Factors and Metric measures
Appendix 10. — Ready Recknoer for Herbicide Solutions
Appendix 11. — Noxious weeds entered into India
Appendix 12. — Requirement of Commercial product of Liquid Herbicides
Appendix 13. — Seasons and its charactiristics
Appendix 14. — World's largest producers of Crops
Biblography 227

1 INTRODUCTION

'And Nature, the old nurse, took
The young child on her knee,
Saying : Here is a story book
Thy father hath written for thee'

Weeds are unwanted plants that interfere with the utilization of terrestrial and aquatic-ecosystems which results in adverse affect on human welfare. They are also referred as plants out of place. Undesirable vegetation grows not only in economic crops but also flowerishes in non-cropped areas such as road-sides, industrial sites, railway lines, air-fields and water ways and causes discomfort to the society. The word *"Weed"* is loaded with values endowed by humans in relation to their own activities. Any plant is considered as economic plant or weed based upon anthro-centric concept. For example, few barley plants in wheat fields are considered as weeds. These plants are termed conditional (relative) weeds. Sometimes, congress grass (*Parthenium hysterophorus*) was found in maize and sorghum fields. These weeds are poisonous and harmful to humans and called as unconditional (absolute) weeds.

1.1 Biological constraints of crop growth

Weeds are the most severe and wide spread biological constraint to crop production. Potential yield reductions caused by uncontrolled weed growth throughout crop season have been estimated to be in the range of 10-90 per cent depending on ecological and climatic conditions. Of the various pests, weeds account for 45 per cent of annual loss of agricultural produce in India whereas insect pests, diseases and other pests causing a loss of 30 per cent, 20 per cent and 5 per cent respectively. In every crop, reduction in biological yield has a direct strong correlation with weed mass. As a thumb rule, *'for every increase in kilogram of weed growth, corresponds to reduction in one kilogram of crop biomass'*

1.2 Expanding behaviour of noxious weeds

Of the 250,000 plants species present on the earth, 1800 weed species causes economic losses in crop production. Nearly 300 weed species have become prominent weeds of the world and causing serious economic losses in cultivated crops. The number of serious weeds are gradually narrowing down as a result of changes in management practices and plant types in crops. At the same time, the noxious weeds which are difficult to control are increasing and expanding their horizon in world agriculture.

1.3 Change in growth behaviour

Every weed has its own growth behaviour and modified with surrounding environment. There are also differences between growth behaviour of annuals and perennials. Annual weeds have a fast and vigorous early growth, produce high proportion of biomass and accumulates in above-ground plant parts. These weeds prepare for generative phase ending in seed setting. On the other hand, perennial weeds are slow at the early start, and invest most of their photosynthates manufactured in leaves, in the underground plant parts like rhizomes, stolons and tubers. They expand their vegetative propagation system slowely but steadily and strongly.

Sometimes annuals behave as perennials and perennials acts as annuals depends upon certain ecological changes. In tropics, there are grey areas in the annual and perennial division of weeds. Some annual weeds behave a perennials if the rainfall is sufficient and uniformely distributed throughout the year. Few perennials complete their life cycle within the year under drought and flood conditions which are not condusive for their growth and development.

1.4 Negative impact of weeds

It is an universal fact that weeds reduce the crop yields by competing for growth factors with crop plants. The intensity of weed competition mainly depends upon nature of weed flora, degree of infestation, life span of weed species, competitive ability of weed and soil and climatic conditions.

Weed seedlings establish in the field very early before the crop seedlings. Early establishment facilitate early colonization and expansion of plant organs. Uncontrolled weeds grow taller than crop plants and curtail the photopynthetic process of crop plants and curtail the photosynthetic process of crop plants and reduce their productivity. It is estimated that 12 per cent of total food production was lossing every year due to interference of weeds. In highly weed infested crop fields, proper weed control may triple the crop yields.

1.5 Weeds as reservoirs of pest and diseases

Herbivorous insects which eat the crop plants originates from non-crop plants in and around the crop fields. Weeds serve as alternate and colletaral hosts to many insect-pests, nematodes and disease caused microbes. The stem flies, thrips, aphids and weevils dwelve on rag weed, pig weed, (*Amaranthus retroflexus*), wild carrot (*Daucus carota*) and wild mustard (*Brassica kaber*) and helps in their multiplication. The insects moving from weeds to crop plants spread the viruses and other pathogens and causes serious reduction in yield. The cost of cultivation of crops also enhanced for controlling the diseases.

1.6 Land values are perished by perennials

In abandoned and neglected lands some weeds extend their life through vegetative propagation and perennials gradually colonized in the natural succession. If the perennials build-up strong infestations it prevents the cultivation of lands. In several states, perennial rhizomatous grasses such as kans (*Saccharam spontaneum*) and thatch grass (*Imperata cylindrica*) infest the lands and make them unfit for cultivation. These infested weed lands require heavy equipment for breaking the sod. Poor peasants have a *desi* plough which is not strong enough of invert the soil clods and expose the rhizomes to sunlight. Kans grass is a mimetic weed in sugarcane fields of Uttar Pradesh, Madhya Pradesh and Andhra Pradesh. Dub grass (*Cynodon dactylon*) is a heavy problem in sorghum fields of peninsular India. *Imperate cylindrica* is a fire-resistant perennial and serve problem in the shifting cultivation cycle. Heavy tractors with soil inverted steel ploughs are required to control them. Marginal and submarginal farmers kept the fields vacant due to lack of capital. The value of these lands rapidly declined and become orphans.

1.7 Inferior quality of produce

The quality of crop produce is usually decline in weed infested fields. Presence of wild onion (*Allium canadense*) in onion fields, wild oats (*Avena fatua*) in wheat crop and *Echinochloa* spp in rice fields greatly reduce to quality of produce. Contamination of noxious weeds in seed grain render them unsaleable. Presence of weeds such as *phaseolus* spp and *mimosae* spp induce offensive odour in dairy products if the cattle are eaten in the weed infested pastures.

1.8 Obstruction of water flows and water regulating structures

In still water bodies aquatic weeds such as *Ipomoea aquatica, Eichhornia crassipes, Hydrilla verticillate, Salvinia molesta* and *Nymphaea stellate* etc. and obstruct the water flows of irrigation canals.

The weeds like *Typha angustata* infests canal banks of drainage channels and streams and causes of floods. Meagre delivery of water in irrigation canals results in inadequate supplies of water to distant fields and tail end areas. Declined flow velocities causes heavy siltation and decrease in carrying capacity of irrigation systems. Water weeds also obstructs irrigation equipment and weirs. The recreational activities such as swimming, fishing and boating is not possible in weed infested water bodies.

1.9 Social issues

Among the most prominent social issues touched with the spread of herbicidal technology in Asian crop production systems and its impact on the displacement of unskilled labour and consequent effect on farm communities. In densely populated rural areas where high levels of unemployment exists, the creation of employment is considered to be the main agricultural policy. Some regions are undergoing rapid growth and in these regions there is high opportunity cost of labour. These labourers are quite happy to discard the arduous task of hand weeding and search for other forms of employment. Crop husbandry should not be high drudgery so that it become less attractive to people. Young people are prefering migrating to cities and metropolis to look for factory employment due to change in attitudes. Once the farmers have adopted herbicidal technology, they do not look back to weeding, even the rates fall significantly.

1.10 Ecological risks

Herbicidal technology usually displace the unskilled labour from the farm. In addition to creation of potential unemployment, the herbicides may pose potential ecological hazard and health damage of humans and animals.

All herbicides are toxic to some degree and it is often main source of problem when handling the chemical. Chronic diseases may develop in users due to continuous use of chemicals. Sometimes profits earned from rice production system that use herbicide may exceed net benefits to society. Society may even suffer a great loss. Recent studies indicates that some herbicides may cause the bronchial asthama, pterygium (eye irritation), poly neuropathy, low haemoglobin, and pulmonary disorders.

1.11 Threat to aquaculture systems

Some herbicides such as butachlor is highly toxic to *tilapia*, a fish widely cultivated under the fish-cum-rice system in some south-east Asian countries. High fish mortality, particularly carps is common in thiobencarb and molinate contaminated ponds. Herbicides can damage the aquaculture systems indirectly. They have a great impact on pH, dissolved oxygen and ammonia in water and accumulate toxic chemicals in the pond sediments. Herbicides poses a danger to consumers through the food chain.

1.12 Herbicides on soil enzymes

Herbicides applied to the rice fields influence soil enzymes considerably which reflects on nutrient solubility indirectly. Research indicates that the soil enzymes such as *amylase, dehydrogenase, urease, invertase* are temporarily inhibited by certain herbicides such as atrazine, 2, 4-D, fluchloralin, butachlor, and oxyfluorfen. High doses of herbicides required to inhibit the enzymatic activity of soil. For example, 10 kg of MCPA ha^{-1} (recommended dose : 2.25 kg/ha) is needed to decrease cellulolytic populations.

In flooded rice fields, application of molinate reduce the decomposition of organic matter and ammonium production. Phosphorus availability will be affected by the capacities of active chemical groups of pesticides to substitute for P ions freed from stable combination of soil constitutents.

1.13 Disturbance of nutrient cycles

In general thiocarbamates (Thiobencarb), amides (Butachlor) and diphenylether group (Nitrofen, Chlomethoxynil and

Chlornitrofen) do not affect ammonification process of soils if applied at recommended rates. Very high rates (15 kg/ha) of thiobencarb sometimes suppress the ammonification significantly. Propanil at 3–5 kg ha^{-1} cause to decline the populations of nitrifiers. Bifenox also inhibit nitrifying activity during first week when applied at recommended level. Negative effects of herbicides are also beneficial because N-losses will be minimum. These chemicals can be used as nitrification inhibitors.

Pesticide application usually do not affects the denitrification process. Denitrifying microflora is a complex and versatile and metabolize a wide range of substrates. Decrease in denitrification is possible when propanil (10.5 kg/ha) and MCPA (2.3 kg/ha) are applied to the rice field. The microbial biomass of soil is regarded as a major channel through which nutrients are transferred to rice. Carbofuron along with nitrofen has a stimulatory effect on biological N-fixing organisms.

1.14 Detrimental impact on microbial metabolism

Application of single herbicide repeatedly may cause changes in the pattern of its metabolic degradation. Such changes in degradation pathways could lead to some soil problems in case of thiobencarb and parathion. Thiobencarb is detoxified by hydrolysis when applied newly. Its repeated application to waterlogged soils favours the multiplication of anaerobic bacteria that degrade thiobencarb by reductive dechlorination. This results in the formations of phytotoxic compounds (*S-benzyl N, N-die ethylthiocarbamate*) that causes dwarfing in rice.

Some herbicides involve in decreasing populations of aquatic oligochaetes and the translocation of the nutrients accumulating at the soil surface to a deeper soil layer. Continuous pesticide application have resulted in declining rice yields over time. Intensive hand weeding, herbicide use combined with a dense rice canopy that could restrict the growth of the photosynthetic aquatic biomass. Shading of paddy fields results in decline in soil microbial biomass to an extent of 20 per cent in a two year period.

2

GENESIS OF SOIL AND XERARCH SUCCESSION

No time to see, when woodlands pass Where squirrels hide their nuts in grass.
—Wordsworth

The parent material of the soil is formed from the weathering of the rocks of earth crust. Minerals are obtained from three types of rocks, namely (a) Igneous, (b) Sedimentary and (c) Metamorphic forms upon weathering.

2.1 Natural weather forces transform the rocks

Natural forces such as temperature, water, ice and wind which are active, play a great role in the transformation of size and shape of rocks. The rocks have been decaying continuously due to heaty days and cool nights. Cracks will be formed due to expansion and contraction of minerals present in the rock because of fluctuations of environmental temperatures.

Water also plays significant role in the genesis of soil. When water flows down speedily it carried many minerals and tiny rocks along with it. As the water possesses enormous erosive power, huge quantities of material will be carried away from one place to another. Large valleys and ravines are found at various places of the world are the result of water erosion. Sometimes, deltas of rivers are disappeared and deposited in the lakes and seas in the end. Ice has tremendous capacity of erosion and acts as effective transporting agency of minerals. Rocks get worn down by moving glaciers.

High speed winds always been an important carrying capacity. When armed with fine sand, it exerts an abrasive action also. Dust storms of high magnitude may occur in deserts that much of the material in filched from one place and transferred to another end. In arid lands, wind is the primary agency of rock weathering.

2.2 Chemical weathering

Chemical process are responsible for the chemical weathering. The rocks undergo several chemical process such as oxidation, reduction, hydration, hydrolysis, de-silicification, De-alkalization, carbonation and solution. The rocks like microcline, limestone, gypsum and those with calcareous content get weathered by the solvent action of water. The presence of carbon dioxide greatly increases the solvent power of water.

2.3 Biological weathering

Biological weathering involves both physical and chemical weathering by micro and macro plants. The lower forms of plants such as bacteria, fungi and lichens begin to gain a foothold.

2.3.1 *Pioneer stage : crustose lichens colonise the rocks*

The pioneer colonisers on the base rock area are blue-green algae and lichens. The mucilaginous algae like *Scytonema* is found to adhere to the rock by it's cell walls. This algae tolerates extremes of temperatures and fix the nitrogen from the atmosphere. Crustose lichens such as *Lacidia, Rinodina, Leconara* and *rhizocarpus* are initial colonizers. Lichens slowly exert its effect upon rocks. Small amounts of soil accumulates under the clusters of lichens. These lichens form a membranous crust. *Crustose lichens* usually grow only when the water is available in the habitat. In hot summers, they appear dessiccated but remain alive. Their spongy nature enable them to absorb water from a small rain, dew and atmospheric humidity. These lichens propagate through their enormous number of spores and soridia. Their dispersal is carried out through wind and water.

These lichens liberates CO_2 excessively during the respiration and helps in the formation of carbonic acid. Their dead bodies also add organic substances to the soil. The decayed lichens along with corroded rock particles make a thin coat of soil on bare rock surface.

2.3.2 *Foliose - Lichen stage : Fructicose lichens flourish*

After the formation of small quantity of soil and humus, the rock surface is less exposed to weather parameters. Nitrogen compounds formed from atmosphere in the lightening of thunderstorms are brought to the soil by precipitation. On the more weathered depressions and crevices of rock xeric foliose and fructicose lichens like *Dematocarpon, parmelia,* and *Umbellicaria* flourish. These foliose lichens gradually replace previously occupied crustose forms.

Foliose lichens possess delicate leaf like thalli that cover the rocks and over shadow the preexisting crustose forms. When the supply the radiant energy cut off, the crustose forms begin to die and decay. Foliose lichens absorb and accumulate minerals and water and arrest the evaporation from soil surface. Humus is accumulated speedily due to decay of lichen bodies. The rocks are loosened into small particles and pulverized due to secretion of carbonic acid from the roots of foliose forms. The water holding and nutrient retention capacity of the habitat is increased with further deposition of soil particles and humus.

2.3.3 *Moss stage : mosses overshadow the lichens*

When the enough humus, dust and rock particles and moisture accumulated in the crevices and depressions of rocks, *Xerophytic masses* appear to grow. Most commonly three important species such as *Grimmia* (black moss) *Tortula* (Troisted mass) and *polytrichum* (hair mass) invade the sites. These mosses are also highly adapted to grow under drought conditions. Gradually the conditions become less favourable for the existing *foliose lichens* and they so slowly start disappearing. The mosses develop rhizoids that penetrate deep into the rocky soil. Sometimes lichens and mosses work together in the weathering process. After several years, the mosses compete vigorously for nutrients and water with the opponent litchens. The erect shoots of mosses overshadow the lichens and eliminate them. The decaying older parts and older mosses form a thick mat over a rock surface which can hold huge quantities of water. This wet habitat is more suitable to herbs other than mosses. Consequently herbaceous community occupied the habitat by replacing mass community.

2.3.4 *Herbaceous stage : Annuals replaced by perennials*

As the soil thickness enhances, the herbaceous vegetation primarily consists of annual weeds make their appearance. This short lived annual grasses and sedges are replaced by biennials gradually and later invaded by perennials like thatch grass and xerophytic grasses.

The roots of herbaceous plants penetrate down upto the bedrock where they produce carbon dioxide and organic acids and accelerates the disintegration of rocks. The carbon dioxide dissolve in water and form into carbonic acid which dissolve minerals present in the rock

$$CO_2 + H_2O \longrightarrow H_2CO_3$$

The type of root development and the return of vegetable matter to the surface soil by plants play an active role in the nutrient cycle. Accumulation of humus is significantly enhanced as the tangled network of root increases. Rapid breaking of rocks enhance the thickness of soil. The evaporations and temperature of soil is decreased as the soil shadowed. This will shorten the drought periods owing the increased capacity of soil moisture.

2.3.5 *Shrub Stage : Enhancement of macro animal activity in soil*

With the change of habitat, the herbaceous weeds shows the sign of degeneration. Several reactions make the habitat more suitable for woody shrubs than the existing ones. The small roots penetrate rock crevices and split them. The fluids and acids found in the root hair also decompose rocks.

Earthworms, insects and other macro animals play a great role in modifying the physical and chemical properties of the soil. The continuous burrowing of earth worms establishes drainage channels in the soil and changes soil structure. The rats and ants form big deposits on the surface of soil. The dead bodies of all animals add organic matter to the soil.

Earthworms form the most important group of macro animals that influence the soil formation. Their significance in plant growth was first stressed by Charles Darwin and V. Henson. Several species are found in the soil. The bigger earthworms are up to 25 cm in length and weigh 3-7 g while the smaller one are about 2.5 mm long and weighing about 0.05 g. They live under specific habitat conditions. Frost and drought are unbearable to them and therefore they do not appear sufficiently in dry sandy soil and thin soils over lying rocks. The worms require reasonable aeration and hence heavy clays with poor drainage are unfavourable to them. These underground organisms prefer a moist environment, heavy organic matter and enough quantity of available calcium. These worms disappear in acid soils. Where earthworm flourish they dominate the whole soil fauna and constitute up to 75 per cent of total weight of macro-animals present in the soil. In soils rich in humus or grass land soils, their numbers are found to be more than 25 lakhs/ha. It is estimated that they weigh about 200-1000 kg/ha.

In addition to earthworms, larger animals belongs to arthropoda and gasteropoda helps in the activation of pulverisation, granulation and tranlocation of soils. The roots of shrubs reach the surface of unpulverized rocks and corrode huge quantities of rock particles which make the soil more massive. Wind movement get declined and humidity of atmosphere increased relatively. Soil become deep and highly fertile and become condusive for the emergence of tree seedlings in place of shrubs. Xerophytic trees invade the area gradually in the place where the shrubs are established.

2.3.6 *Forest stage : Replacement of xerophytic to mesophytic tree species*

The first established trees are relatively xeric, dwarf and sparcely distributed here and there. Their seedlings are heliophyllous and stunted in growth. With the formation of more organic debris, the water holding capacity is still increased. The spongy soil holds huge quantities of water and arrest the water erosion. The xerophytic trees gradually vanished as the humidity of atmosphere increases. Subsequently mesophytic types develop and increase in number and vigour.

The mesophytic trees are deeply rooted and forage more soil for nutrients and water. They grow densely and become dominant because their seedings are shade tolerants. Under the shade of large mesophytic trees, some shade loving shrubs and herbs starts to grow. These new herbacious plants are well adopted to damp weather and highly fertile soil. They form their own communities which are relatively stable. After a very long interval, complete harmony develops between plant communities and environment. This climax forest remains unchanged unless some significant environmental changes disturbs it.

In this way, the succession of given type of vegetation is not a sudden, hapazard covering of land. It is an orderly process of colonization of plants. This process takes place several hundred years and governed by both environmental factors and plant species. Succession is the interaction of both.

3

SOIL—A PROMINENT ABIOTIC EDAPHIC FACTOR

The most tiny weed emerges from the earth
Can give thoughts that often lie in the ocean deeps.
—French Maxim

The soil is an example of a major ecosystem, containing a huge number of different kinds of macro and micro-fauna, micro-flora and plants and that compose an inter-related biological complex. Soil is a complex physical and biological system providing support, water, nutrients and oxygen for the organisms. In another words, soil can be termed as a bridge between the inorganic, organic and living worlds. Indian soils can be divided into four major groups.

3.1 Indo-Gangetic alluvium

3.1.1 *Formation*

These soils have been formed from the sediments deposited by the Indus, the Ganges, the Brahmaputra and their tributaries. These rivers while flowing down the Himalayas bring with them the products of weathered rocks constituting the mountains, in varying degree of fineness and deposit them while they traverse the plains.

3.1.2 *Occurrence*

The alluvial soils are found in Punjab, U.P., Bihar, West Bengal, Orissa and Assam. These soils occupy an area of about 7 lakh square kilometres and comprise biggest areas of Indian soils. Nearly 42 per cent of the total population of India lives on it. All most all crops can be grown in these soils but wheat, maize, rice, sugarcane, gram, sorghum and pearl millet give luxuriant yields. In another words, these soils can be termed as best soils of India and cannot be compared with others.

3.1.3 *Characteristics*

- These are loamy in texture, lacking in cementing material and loose in arrangement.
- Can be worked easily.
- Conspicuous response to irrigation and manuring.
- Gully erosion is very serious in high rainfall areas.
- In some of the deep clays, impervious *kankar* layers are formed at certain depths which results in obstruction of drainage.
- The soils contain sufficient quantity of lime, less nitrogen and organic matter. Potassium is sufficient in these soils but poor in phosphorus.
- They do not show any profile characteristic. At some places, lower layers contain *kankar* nodules.

3.2 Black cotton or Regur soils

3.2.1 *Formation*

As the name indicated, the soils are black in colour. The soil colour does not change up to a depth of 2-4 metres. These soils are formed in the broken country, between the hills and plains. They are constantly enriched by the the additions washed down from the mountains and hills. The black colour is formed due to presence of *Titaniferrous magnetite*. Some people believe that black colour is attributed due to presence of 1-2 per cent organic matter in the soils. Presence of hydrated silicates of iron and aluminium may also induce black colour to the *regur* soils. Most of the black soils derived from Deccan trap.

3.2.2 *Occurrence*

The soils are mostly found in Maharashtra, Madhya Pradesh, Andhra Pradesh and Gujarat. Few patches are found in Tamil Nadu and Rajasthan. Black soils cover an area of about 3 lakh square kilometres. Maharashtra occupies larger part of the area under this group. On slopes, these soils are light coloured, shallow and poor in fertility, while in valleys, they are black and deep. In the valleys of western ghats the soil often 6-8 m depth. The soils have a high degree of fertility and compared with the productivity of *prairies* of U.S.A. and *chernozems* of Soviet Russia. Some of the soils have high sustainability index because they are fairly productive even after a period of two centuries with little fertilization. Regur soils are usually derived from ferromagnesian schits in particular a layer rich in *kankar* nodules is formed. This layer is formed due to segregation of calcium carbonate in the underground soil.

As the name implies cotton is the important crop grown on these soils. Besides this, rice, soybean, sorghum, pearl millet, wheat, bangal gram also gives bumper yields. Some crops like tobacco, chillies, garlic and onion are grown under rainfed conditions.

3.2.3 *Characteristics*

- These soils are highly argillaceous and have a tendency of adhesive when wet conditions. They develop deep cracks during summer and called as self-tilled soils.
- Possess high water holding and retentive power of moisture.
- Contains large quantity of calcium and magnesium carbonates.
- Possess fairly high quantity of iron, lime and alumina.

- Potassium content varies with parent material.
- Poor in phosphorus, nitrogen and organic carbon.
- Rich in montmorillonite and beidellitic group of minerals and high in cation exchange capacity i.e. 40-60 m. eq.
- Subjected to sheet type of soil erosion in heavy rainfall areas.
- Show good response to irrigation but field should be laid out in ridge and furrow method to avoid water logging.
- Irrigation may cause the accumulation of soluble salts on the surface and therefore drainage is essential in these soils for keeping the productivity of soils.
- Clay content of these soils are usually varies from 35-50 per cent and O_2/R_2O_3 varies from 3 to 3.5.
- In South India, some soils contain about 65-80 per cent clay fractions and high values of specific gravity, pore space and hygroscopic coefficient. These soils have high pH and rich in lime.
- *Clay soil* usually contains more than 30 per cent clay where as *sandy soil* contains 10 per cent each of clay and silt with a large proportion of sand. In case of *loamy soil* there are about 30 per cent each of clay and silt, the rest being sand.

3.3 Red Soils

3.3.1 *Formation*

Red soils also comprise an important group of soils. They are derived from the material obtained by the weathering of the ancient crystalline and metamorphic rocks. The red colour is mainly due to presence of iron oxides in the soil. The soils are termed as sedentary soils as they formed with weathering materials. Many soils which have no red colour also come under this group. Some alluvial soils are also included in it. Certain lateritic soils, though they are reddish in colour but not come under the red soils. The clay fraction of the red soils is rich in Kaolinitic type of minerals. Most of the red soils have been found under forest vegetation.

3.3.2 *Occurrence*

Red soils occupy largest area in South India constituting two- thirds of the cultivated area. They occur in Andhra Pradesh, Tamil Nadu, Karnataka, Bihar, Uttar Pradesh and Madhya Pradesh. The red soils of South India are shallow with open texture. In Bihar, these soils are found in the districts of Hazaribagh, Santhal Paraganas, Singbhum and Mungbhum. These are also occurred in the districts of Bundelkhand, Jhansi and Hamirpur in Uttar Pradesh.

3.3.3 *Characteristics*

- The soils contain sandy clay in texture, coloured red by iron peroxide.
- Morphologically these soils can be divided into red loams and red earths.
- Red loams are of argillaceous nature with clody structure and low in concretional material.
- Red earths are rich in concretion with loose and friable surface soil.
- Their base-status is low and pH ranges from 6.6–8.0
- These soils are poor in organic matter (0.1 per cent) and phosphorus.
- The compounds of iron and aluminium are found in high proportions.
- Potassium is adequate in these soils.

3.4 Lateritic soils

3.4.1 *Formation*

These laterites are formed from the weathering of several types of rocks. Under humid and high rainfall areas, the siliceous matter of rocks is leached away to a great extent during weathering. Alternate wet and dry conditions also encourage the formations of these soils.

3.4.2 *Occurrence*

Laterites are found on the hills of Eastern ghats, parts of Assam, South Karnataka, Madhya Pradesh, Orissa, Maharashtra, and Kerala.

3.4.3 *Characteristics*

- Laterites are rich in organic matter and phosphorus.
- Phosphorus is present probably in the form of iron phosphate.
- They are poor in lime, magnesia, nitrogen and potassium.
- Laterites of South India are richer in nutrients, organic matter but low in pH. Paddy is grown in lowland laterites. In uplands plateaus, crops like tea, rubber, coffee and cinchona can be grown.
- Laterites have coarser materials. The ratio of silica to alumina is found higher in their clay fraction.
- They are poor in lime, phosphorus but moderate in organic matter, nitrogen and potassium.
- The Bengal laterites contains low quantity of phosphorus and potassium.
- The ratio of silica to alumina is higher in clay fraction.

3.5 Peaty and Marshy soils

3.5.1 *Formation*

In humid and high rainfall areas, peaty soils are formed due to huge accumulation of organic debris. Decomposition is very slow due to anaerobic conditions.

3.5.2 *Occurrence*

They are largely found in Travencore and Cochin regions of Kerala. They occupy about 75 to 90 square kilometres.

3.5.3 *Characteristics*

- They contain sufficient quantity of soluble salts.
- Peaty soils combined with salinity is called *kari.*
- The soils are submerged during monsoon season and put under rice cultivation.
- The soils are black in colour and highly acidic.
- The pH may go down up to 3.9.
- The acidity is attributed to the formation of organic acids during anaerobic decomposition.
- In some areas, sulphates of iron and aluminium also occur.

3.6 Desert soils

3.6.1 *Formation*

They are formed through strong winds. The sands are largely blown in from the coastal regions and Indus valley.

3.6.2 *Occurrence*

The soils are found in arid regions of Rajasthan, parts of Punjab and Gujarat. The area under these soils is near about 60,000 square kilometres. Though this area lies in the track of South-West monsoon but receives no rain due to peculiar physiographic features.

3.6.3 *Characteristics*

- Desert soils are light in texture and poor in organic matter and nitrogen.

- They possess high pH due to presence of more soluble salts.
- They contain varying quantities of calcium carbonate.
- These soils are not only thirsty but also hungry.
- These soils can brought under cultivation if irrigation is provided.

3.7 Forest and hill soils

3.7.1 *Formation*

The forest soils have been formed by the decomposition of growing vegetation and fallen leaves of trees. The formation of the soils depends upon the vegetation of a particular forest and climate of the area.

3.7.2 *Occurrence*

These soils occupy about 14 per cent of total area. There are two types of soils i.e. *acid humus* and *basic humus*. Soils that formed under acidic condition and contains high quantity of acid humus and less bases.

The soils formed under neutral conditions contains higher proportion of bases is called basic humus soils. Sometimes laterites are also formed from these soils. These soils are found in Assam, U.P., M.P., Bihar and Maharashtra.

3.7.3 *Characteristics*

- These soils are highly fertile and contains rich in organic matter.
- The hilly soils of Assam contains sufficient quantity of organic matter and nitrogen.
- In sub-Himalayan tracts, 4 groups of soils are found viz. Brown forest soils, Padzols, Red loams and Transitional Podzol.

3.8 Saline and alkali soils

3.8.1 *Formation*

Saline and alkali soils are formed due to accumulation soluble salts at the soil surface in arid and semi-arid regions. Alluvial clay and silt may contain undecomposed minerals which on weathering liberate sodium, calcium and magnesium salts. When thee soils are deposited in large quantities such saline soils are formed. The capillary movement of heavy soils may bring the soluble salts of the deeper layers to surface layer.

3.8.2 *Occurrence*

These soils are found throughout India in all climate zones but are very common in arid regions. It is estimated that one lakh hectares in U.P. and 3 lakh hectares of land in Punjab and Haryana have been affected by *user* and 10,000 hectares are being added every year in Punjab.

3.8.3 *Characteristics*

- Saline soils are slightly alkaline because their exchangeable cations are mainly dominated by calcium and magnesium with relatively low quantity of sodium.
- The percentage of exchangeable Na is less than 15 and pH is 8.5 or less.
- No dispersion of granules and colloids are highly floculated.
- Crops can be produced if drainage is installed.
- In alkali soils, sodium hydroxide converted into sodium carbonates and makes the soil solution strongly alkaline.
- Alkalinity causes soil colloids to defloculate or disperse thereby become impervious to drainage.
- Alkali soils contain more than 15 per cent of adsorbed Na.
- The pH lies between 8.5 to 10.0.

3.9 Soil water—a crucial ecological factor

Water is quite essential to all living organisms. Depending upon the water requirements, the weeds are divided ecologically into *hydrophytes, mesophytes* and *xerophytes*. Thus the amount of water determines the type of weed flora in a particular habitat and this in turn determines the type of flora and fauna of that habitat. Soil moisture can be divided into five types i.e.

I Groundwater

II Gravitational water

III Capillary water

IV Hygroscopic water

V Combined water

3.9.1 *Groundwater*

It appears below the water-lable. The water moves down the soil and drains out of it and present in the large soil spaces ecologically important in the leaching of nutrients. The depth of water-table varies from place to place. At the same place, its depth changes according to season. The excess water used by plants or evaporated from the surface ultimately goes down through the soil to the groundwater level. If the water-lable is low it can have little effect on moisture condition of the soil but in case of high water-lable, the plants suffers due to lack of aeration and oxygen supply.

3.9.2 *Gravitational water*

It is the free water that moves down into the soil due to action of gravity, until it reaches the water level. This water is loosely held in macropores and moves down to the lower layers under the pull of gravity is known as gravitational water. The tension of the gravitational water is one-third of an atmosphere. This water occurs in the larger pore spaces of the soil, therefore it excludes air from there and interferes with oxidation. Hence, drainage is most essential. At the same time, excess drainage causes loss of nutrients from soil. A soil with poor drainage is not a good medium for producing most of the field crops. Lack of aeration in soil results in slow down of biological and chemical activity and normal soil process. Root growth and mineral uptake are weakened due to anoxic conditions.

3.9.3 *Capillary water*

It is the amount of water held by capillary forces in smaller soil channels when the groundwater has drained. This water is present as this films around the soil particles and granules, either by surface tensions or by interfacial forces of attraction. Capillary water held in the soil between the field capacity and hygroscopic coefficient. Capillary movement is very sluggish between PF 4.2 and 4.5 but ready movement occurs when PF is 2.7. It exists in the liquid state and contains solutes. It remains in the soil for any length of time and functions physically and chemically and is called as soil solution. The percentage capillary water held by different soils is affected by texture, structure and organic matter of the soil and by its position in the soil column.

The capillary capacity of a soil likely increases with the fineness of texture. This is mainly owing to the porosity of minute interstices that carry water films, when the texture of the soil is very fine. In granular soils, the interstitial spaces are enlarged which results in increase in water holding capacity greatly. Because much of the capillary water is kept by such pores, an increase in the retention of water by capillary may occur. The compaction of heavy soils reduces the capillary water held by them while the opposite is true in light textured soils. Humus had very high capillary capacity which is several times greater than minerals. The quantity of capillary water held by a soil varies with the length of the column, the pull of gravity causing water to move towards its base when the tension of PF is zero.

3.9.4 *Hygroscopic water*

When the dry soil is kept in a humid atmosphere, it absorbs some quantity of moisture. When a soil is dried in air at ordinary temperature it still retains a certain quantity of moisture. The moisture closely surrounds the soil particles and is of little ecological importance. The hygroscopic moisture includes all the moisture present at the hygroscopic coefficient and is held with a tension of 31 atmospheres and greater. It exists as a very thin film at the solid-liquid interfaces of soil particles. In this state the soil particles do not stock one another. It participates to a slight degree into the biological activity. This water can be driven out of soil by heating it at 110°C for 10 hours.

The PF of the soil moisture is 6 at hygroscopic coefficient. The percentage of moisture held by the soil at 50 per cent of relative humidity is called hygroscopic coefficient. Hygroscopic coefficient varies from soil to soil. Sandy loams which contains little organic matter may be as low as 1-2 per cent by weight. In heavy soils like clay loam it varies from 15 to 20 per cent. In case of peat soils its value will reach up to 60-70 per cent.

3.9.5 *Combined water or imbibitional water*

Soil water which is absorbed by soil colloids is called imbibitional water. After absorbing this water the soil coloids begin to swell and soil becomes visibly moist as shown by its darkening in colour. Imbibitional water is available to plants but it is held tightly against mechanical forces. The total amount of combined water in the soil is called the *holard*. The amount of water that plants can absorb out of holard is termed as *chesard*. The amount of water that cannot be absorbed by plant is called *echard*. Water holding capacity is low when the soil particles enclose larger spaces. If the spaces are small, the water holding capacity is high. The soils which have larger spaces is termed as *physically dry soil* because of rapid percolation of water to lower layers and not available to plants. Sometimes, the soil has sufficient amount of water but due to salinity it can not be utilized by plants. These soils are termed as the *physiologically dry soil.*

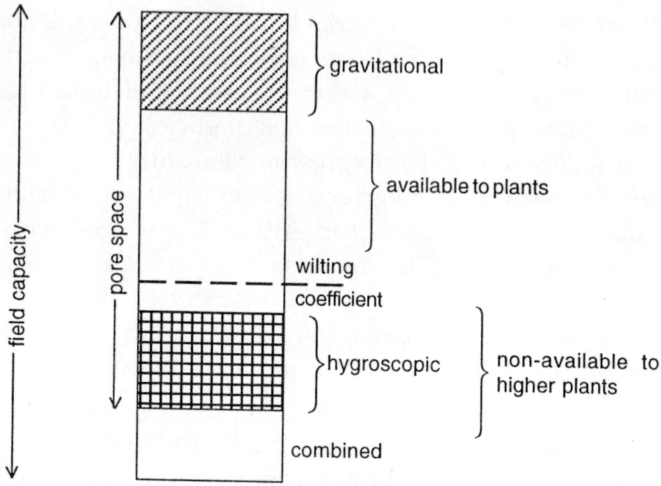

Fig. 3.1 : Type of soil water

3.10 Soil and site description for weed ecological surveys

Geographical situation : Latitude :

Longitude :

Altitude :

Location : District :

Mandal :

Village :

Mean maximum temperature °C :

Mean minimum temperature °C :

Mean Relative humidity (Morning) (%) :

Mean Relative humidity (Afternoon) (%) :

Annual Dew fall (mm) :

Annual Rainfall (mm) :

Rainy days (No.) :

Farming situation : Crop rotation :

Major soil group/series association :

Soil information

Relief : 

Location at the slope : Ridge ( ) , Middle ( ) , Foot ( )

Slope inclination : Slope direction :

Soil moisture : Arid ( ), Dry ( ) , Moist ( ) , Wet ( ), Flooded ( )

Soil structure : Loose ( ), Semiloose ( ), Compact ( ), very compact ( )

Friability : Good ( ), Moderate ( ), bad ( )

Soil texture : Sandy ( ), Sandy loam ( ), Loamy ( ), Clay loam ( ), Clay ( )

Calcium content : No ( ), Poor ( )

(10 per cent HCL)

Colour of the soil : Black, Yellow, Red, Ashy, Brown.

Presence of stones/gravel: Without, low, Medium, high

Depth of soil : Shallow, Medium, deep

Drainage : Good, Moderate, Poor

Soil preparation : Tractor, mould board plough, Desiplough

Irrigation source : Canal, Tank, Borewell, Springs

Weed control method : Manual, Mechanical, Chemical.

4

CHANGE OF WEED FLORA DUE TO AGRO-ECOLOGICAL ENVIRONMENT

As the strong and swift current is a blessing for the swimmer so the hard knocks of life in every living thing are blessing in disguise.

The composition of a crop field's plant community will change with the type of agricultural practices that are applied. Modifying crop management inputs will alter the microenvironment, and the morphological and physiological traits that confer plant success. Agricultural activities influence greatly the structure of weed communities and their degree of competition.

4.1 Responsible factors

- Type of soil or texture
- Land scape position
- Soil-water regime
- Land preparation
- Planting method
- Crop season

- Crop rotation
- Irrigation water
- Herbicide application.
- Nitrogen management

4.1.1 *Soil texture*

Soil texture indirectly influence the soil aeration and water retention capacity. Light textured soils like sandy and sandy loams supports the growth of aerobic weeds like *Celosia argentia, Phyllanthus niruri* and *Mimosa pudica* etc., whereas heavy textured soils encourage the growth of hygrophytic weeds such as *Monochoria vaginalis, Cyperus difformis* and *Ammania baccifera* etc. In general most weed species are tolerant of a wide range of soil types and soil factors are not a major limitation to weed distribution.

4.1.2 *Landscape position*

The composition of weed flora in a particular region is also strongly influenced by landscape position. In coastal areas, the fields have higher ponding potential and favours the growth of aquatic weed.

4.1.3 *Land preparation*

Major changes of weed composition can be seen due to land preparation. Puddled soils have a very few weed species. In these soils *Monochoria vaginalis, Marsilea minuta, Sphenoclea zeylanica, Ludwigia octovalvis* and *Ammania baccifera* become predominant. In dry ploughing areas, several mesophytic weed species like *E.colona, Ageratum conyzoides, Commelina jacobi, Mimosa pudica* and *Leptochloa Chinensis* are increased. Perennial weeds like *Cyperus rotundus* and *Imperata cylindrica* may dominate after few years in the the zero tillage or minimum tillage system.

4.1.4 *Planting method*

Changes in planting method may be one of the most important ecological forces influencing changes in weed dominance and competition in rice fields. Switching from transplanting to direct seeding causes a shift in the dominant broad leaf weeds to grassy weeds. Weed occurrence in dry seeded rice is usually three times higher than transplanting method. *Cyperus difformis* and *E. crus-galli* are the dominant weeds in direct seeded rice whereas *Scirpus juncoides* and *E. crus-galli* are major weeds in wet-seeded rice. Transplanting with infant seedlings encourage more weed growth than using older seedings. Weed growth is usually more in machine-transplanted than hand-transplanted rice.

4.1.5 *Crop season*

Weed community distribution is determined by variation in environmental factors and correlated well with meteorological conditions such as temperature, rainfall, and humidity in the air. The grasses such as *Paspalum paspalodes, Cynodon dactylon* and *E.chinochloa colona* dominate in dry season whereas *Monochoria vaginalis, Marsilea quadrifolia, Ipomoea aquatica,* and *Ludwigia octovalris* mostly appears in rainy season rice, while species like *Hydrilla Verticillata* (L.F.) Royal, *Ottelia alismoides* (L) vahl and *Nymphoides indica* (L) O.K. are found in deep water areas. In lowlands, the water table is higher and encourage the growth of *Ipomoea aquatica, Nelumbo nucifera* and *Blyxa auberti* etc. The lands of low ponding potential and lower water-table encourage the proliferation of hygrophytic weeds such as *Cyperus iria, C. babakan, Scirpus supinus, Molochia concatenata* and *E.chinochloa oryzicola*. The weed species *E. crus-galli* and *Digitaria ciliaris* usually dominates in the mesophytic environment of inland soils.

4.1.6 *Soil-water regime*

Water management practices which are adopted in the cropped fields also greatly influence the weed composition. In continuous flooded rice weeds like *Ludwigia perennis* and *Ammania baccifera* dominates the habitat than other weeds. The weeds *Sphaeranthus indicus*, and *Caesulia axilaris* are very common in semi-water-logged conditions. If the fields are dried, the mesophytes such as *E-chinochloa colona*, *Eclipta prostrata*, *Commelina jacobi*, *Ageratum conyzoides* and *Rumex dentatus* proliferates rapidly. Field capacity is ideal for the germination of *Leptochloa chinensis*, *E. colona* and *Eclipta alba*.

Irrigation intervals also influence the type of weed flora. When sprouted rice is sown on puddled soil, with the field flooded one week after, grases and sedges are usually predominate. If the water is introduced two weeks later, sedges dominate and accounts about 80 per cent of total weed biomass.

4.1.7 *Crop rotation*

Crop rotation is an effective method of minimizing the damage from weeds. Rotation of different crops on the same land has been an important method of weed control. In fact, certain weeds are often associated with specific crops due to similar physiological adoptability. Rotating crops having drastically dissimilar life cycles or cultural conditions break the cycle of weeds and reduce their populations over a time. Intensive cropping systems can increase the competitive ability of crops, thereby reducing weed pressure.

Rotating wetland rice with a dryland crop results in reduced infestation of water tolerant weeds in the rice crop. The upland crops like maize, soybean, groundnut, mungbean when rotated with transplanted rice decrease the mesophytic weeds like *Agrophyron repens*, *Paspalum* spp. and *Imperata cylindrica* etc., weeds in rice increase as the length of time after reconversion to wet land field increases. That is why reconverted wetland field may be returned to a dryland field within a certain time.

Different cropping, soil, water and weed control practices bring about changes in the weed community. For example, *Scirpus maritimus* persists in continuous transplanted rice. When dryland crops are rotated with wetland rice, weed menace will be declined. The build up of **difficult-to-control weeds** are usually occurred in the monoculture systems. Sometimes changes in cultural practices may generate new weed problems. With proper crop rotation, several noxious weeds are discouraged and diverse weed flora is promoted.

4.1.8 *Irrigation water*

Irrigation water is main culprit of spreading of weed seed and vegetative propagules. Billions of seeds are carried out through irrigation water from one place to another. The quantity of moving seeds mainly depends upon their size and weight. The weeds present in irrigation canals and on bunds are the major source of seed. That is why the weeds may be destroyed with chemical or burning before blooming. Shallow or zero water depth in rice usually help in prospering of *E. crus-galli* and *C. difforms*.

4.1.9 *Herbicide application*

Herbicides are the main method of weed control even in Asian countries. Until the mid-1960 2, 4-D and PCP are used in rice. Nitrofen is also widely used in early 1970s. Acid amide-type herbicides like alachlor and butachlor are used more than 20 years, until 1990. Owing to use of acid amide-type and carbonate herbicides there was a marked reduction in grassy weeds such as *Echinochloa* spp. and *Rotala indica*. However, the broad leaved weeds like *Monochoria vaginalis* and *Massilea minuta* have dominated in most of the Asian rice fields due to their continuous use. Continuous use of phynoxy herbicides usually encourage the emergence of grasses like *Echinochloa* spp. and *Leptochloa chinensis*.

Recent use of sulfonyl ureas such as bensulfuron and pyrazosulfuron results in dominance of troublesome weeds like *Sagittaria triforia* and *Echinochloa kuroguwai* over period due to lack of residual effect. Sulfonyl Ureas are for post-emergence broadleaf now-used sedge control and aryloxy phenoxy propionates or APP's (fenoxaprop, cyhalofop) for post-emergence grass control in Asian rice fields.

4.1.10 *Nitrogen Management*

In most of the Asian countries, the recommended dose of nitrogen varies with season, soil and varietal characteristics. The N fertilizer rate ranging from 40 to 150 kg N/ha. However the dry matter production was optimum at 75-100 kg N/ha. Always nitrogen should be top dressed after weeding operation only for effective utilization. At early stage of rice (active tillering) if the N is top dressed without weed control, the major share of fertilizer will be taken away by weeds. Weeds are over topped the crop plants due to utilization of applied nitrogen. The recommended practice is a three-split application i.e., 1/3rd basal + 1/3 at active tillering +1/3, 5 days before panicle initiation.

If the weeds are not removed, it is best not to apply or postpone the application of N until the rate of N absorption has slowed down. The N uptake will be decreased after the weeds have flowered. Weeds are strong competitors with rice both at low and high N levels. At higher rates, weed growth is rapid, resulting in greater weed competitors and lower rice yields than non-fertilized plots.

4.2 Weed-competing ability of cultivars

Modern rice cultivars cannot compete for resources with weeds due to their short stature and erect leaves. Tall, fast growing, droopy leaves, fast canopy forming, and high tillering capacity are believed to be ideal characteristics that contributes better weed-competing ability. However, these traits are quite contrasting that favour higher yields at high fertilizer inputs. Rice breeding programmes are not designed to compete with weeds so far, with an assumption that weed control imputs are easily available.

The existing intermediate statured rices are easily outgrown by tall, fast-growing grass weeds like *Echinochloa crus-galli* and *E. glabrescenca* in irrigated rice and *Rottboellia cochinchinensis* in dryland rice. The short rices do not have the head start over weeds in both direct seeded and transplanted conditions. The rice plants are highly susceptible within the first 30 days of the life cycle for the weed competition.

4.3 Morphophysiological traits to improve the competitive ability with weeds

- The presence of chlorophyll in the first leaf like barnyard grass.
- This chlorophyll helps in a 4-day head start over weed seedlings.
- The higher photosynthetic ability that can utilize minerals and water.
- High potential of acclimatization for various situations.
- Exploration of allelopathic cultivars.
- Presence of adequate carbohydrates in the endosperm.

4.4 Seedling rate

Seedling rate should always be optimum to get higher per unit area yields. Low seedling density gives opportunity to weed germination and growth. All open spaces should be occupied by rice seedlings to prevent weed growth. Very high density or high seed rates suppress the weed growth effectively but results in intraspecific competition. This competition leads to poor crop yields. When different factors are taken into account, a seeding rate of 100 kg ha^{-1} is found to be good, in view of cost-benefit ratio in rice. In irrigated rice, a spacing of 20 × 15 cm in rainy season and 15 × 15 cm during winter are found to be optimum to get higher yields. When weed pressures are intense, decreased spacing may work in semidwarf rices. Optimum seeding rate is highly critical in areas where weed control inputs are scarce. There is no need of high seedling rates if control measures are easily available and adequate enough.

4.5 Prevention is always better

Certain measures to avoid weed seed dispersal or build up of weed seed bank in the field is thousands times better than sophisticated control measures. Using weed-free seeds, weed seedling-free crop seedlings cut the weed infestation at initial stage. Several weed seeds spread through contaminated farm equipment like tractors, puddlers, harvesters and combiners. Farm equipment should be clean and should be weed free when going to new fields. Maintenance of clean field borders, levies, irrigation canals and cultivated fields are also essential to stop weed dispersal. Weed infestation during fallow periods should not be neglected as it will build-up seed banks in the soil.

The clean and certified seed is the only answer for preventing the weed seeds entering into crop fields. In several tropical Asian countries farmers use the seed from previous harvests to lower the production costs. Several grass seedlings which are difficult to distinguish from rice seedlings in nursery are transplanted in the main field. Several farmers do not show equal attention in maintaining clean fields during fallow periods. Weeds allowed to grow and set seeds between cropping seasons contributes to the additional seed reserves in the field. The non-selective herbicides like paraquat, glufosinate and glyphosate should be sprayed on weeds that grow after crop harvests. Chinese farmers usually burn the rice straw and crop residues after crop harvest to kill the existing weeds.

4.6 Ideal morpho-physiological traits of rice for weed competition

Heavy weed infestation is considered as one of the major constraint for successful rice production. Scarcity of agricultural labour is increasing day by day in all Asian nations. Many farmers in a number of countries are switching from hand weeding to chemical weed control due to hike of labour wages. Increasing dependence on heavy use of chemicals for weed control leads to several problems like environmental pollution and swifting of weed flora. Hitherto varietal improvement in cereals are concentrated for pest and disease resistance besides higher yields. Attention was not paid to improve morphological traits of rice, to withstand or overcome the weed competition. This attempt is worth exploring because it leads to reduced herbicide application. The morphological characteristics for successful weed competition include early vigour, rapid seedling emergence under oxygen free environment, smothering canopy, tolerance for parasitic weeds, allelopathy and resistance to herbicide.

4.7 Long-duration varieties count down the weed competition

Several studies indicates that long-duration cultivars can withstand the competition posed by barnyard grasses (*Echinochloa* spp.). Yield losses due to wed infestation are greater in short-duration varieties than long-duration rices, even though they had the same height. Short duration rices (100-120 days) do not have opportunity to recover from grass weed Competition. Several *Echinochloa* spp. mature much earlier than short duration rice. The longer field duration compensate from competition from weeds.

4.8 Tallness indicates high competing ability

Taller rices have higher competeting ability with weeds and smother surrounding weeds. Plant height is highly correlated with weed suppressing ability of crop. Yield reduction due to weeds is less in case of taller varieties. Short statured varieties cannot utilize above ground growth requirement like solar radiation effectively.

4.9 Longer, droopy leaves weaken the weed growth

Cultivars with long, droopy leaves are more competitive than those with short, erect leaves. Strong competitors have more leaves than weak competitors. Long, droopy leaves shadow the surrounding weeds and weaken their photosynthetic activity. Droopy leaves forms a wide angle with the stem which is advantageous in competing with noxious weeds. That is why, tall traditional cultivars smother the weed growth successfully.

4.10 Profuse tillering capacity

High tillering capacity is closely correlated with strong competitive ability. High tillering cultivars develop their leaf canopy faster and crowd out weeds. Shy tillered varieties creates gaps in the field and provide niches to weed emergence.

4.11 Greater root volume

Roots are the foundation of plants but studies are negligible than above ground plant organs. Varieties differ as much in the root system as in parts present above the ground. Varieties vary in root elongation, root branching, overall length of roots, diameter of roots and occupational soil volume. A strong root system will withstand competition with other cultivars, intercropped species and weeds. More rapidly produced root system, a more extensive and strategically placed root system, rapid uptake of minerals and water and greater drought tolerance are essential characters required for competing for substrate resources.

4.12 Seedling survival under anoxic conditions

In several countries, sowing pregerminated seeds into standing water of puddled soil. Cultivars should have their ability to emerge through standing water. Such type of characteristic is effective for weed control because several grasses and broad leaved weeds cannot emerge through soil, covered with water. In temperate countries, water temperatures are low and oxygen levels are also high and sustain seedling growth and emergence of rice. In contrast, the amount of dissolved oxygen is very low and water temperatures are high in tropics and subtropics where rice cultivars fails to emerge from anaerobic soil environment. Adoption of seeding in flood water using varieties tolerant of a low oxygen soil environment helps in reducing the herbicide use for weed control.

4.13 Inducing resistance for parasitic weeds

Some of the parasitic weeds like *Striga* spp. parasitise on dryland rice and absorb nutrients from the post plant and causes severe yield losses. In some of Asian and African countries, the yield loses due to *striga* goes to an extent of 95 per cent. In some areas of Tamil Nadu and Kerala, *striga* infestation was so severe that farmers were forced to grow another crops or to abandon growing upland rice. Some Kenyan upland rice lines were found to be resistant to *striga* infestation. The varieties Ble Chai, B 3913 F-16-5-St-4, IR 38547-B-B-B-7-2-2 and IR 47255-B-B-5-4 are also found to be resistant to moderate infestations.

4.14 Herbicide resistance

All plants and crops have some natural tolerance for one or more herbicides. This innate natural tolerance can be exploited in developing herbicide resistance. Modern biological tools shows the way to develop varieties resistant to can be obtained from the crop gene pool or from bacteria like organism. Some clonal variants can also be selected in tissue culture. Recently bacterial genes are introduced into rice for herbicidal resistance particularly to bialaphos and glufosinate-ammonium. The progeny of transgenic plants carrying this gene is having higher resistance even at 500 ppm of bialaphos. The resistant genes can also be transferred to closely related wild species of rice by outcrossing in areas where wild and cultivated rice are sympatric.

4.15 Plant allelochemicals

The plant allelopathic chemicals are secondary metabolites that play either beneficial or detrimental, in plant-plant, plant-microbes, and plant-weed interactions. The interactions play great role in plant dominance, succession and crop productivity. The allelochemicals, may be released into the environment from plants by means of leaching volatilization, root exudation and decomposition of plant residues in the soil. Some rice genotypes have allelochemicals, which serve as weed control agents. If these are introduced into the present cultivars, it reduces the need of herbicide chemicals.

Improved Japonica cultivars have little allelochemicals but traditional tropical Japonicas and red rice strains have strong allelopathic activity.

Rice allelopathic chemicals may not be effective against all types of rice weeds. Even weeds also produce allelochemicals and to suppress other co-weeds. These chemicals can be cloned and introduced through genetic transformation in rice leads to less investment on weed control. However, that over time, weeds may develop resistance to allelochemicals. With that result, new weed flora may dominant in rice fields due to change in ecology.

5

SHIFTS IN WEED FLORA

"Struggle is never useless though wounds are painful'.

The meaning of weed control does not mean simply removing one or two species from within the plant community. This modifies the relationships among constituent species. Some weed control measures usually disturb the micro-ecosystem, altering competitive interactions, creating stress to one component and disrupts the natural pattern of development. As a consequence, weeds which are of minor importance may emerge as primary problems.

5.1 Change in cropping system

When a cropping system changes, the entire microenvironment may alter which may cause unintentional changes in weed flora composition in the same field. Several factors like light, soil water and nutrients may contribute to the process of shifting of weed species. Adoption of new agronomic practices may sometimes be conducive in creating opportunities for imported weed seeds or secondary important weeds. These may create major infestations in that particular cropping system.

5.2 Shifting of weed flora in dwarf wheats

Grassy weeds were never considered a serious problem in traditional tall wheat cultivation of India. After the advent of mexican dwarf and semi-dwarf wheats, it is believed to have generated serious new weed problems in the form of canary grass (*Phalaris minor*) and wild oats (*Avena ludoviciana*). Surprizingly none of these weed species was a problem in tall wheats. It is believed that some ecological conditions appear to have favoured the growth and dissemination of these weeds.

Along with new dwarf wheats, the whole package of new cultivation practices were also adopted by progressive farming community. More irrigations and heavy N P K fertilization changed the crop ecosystem and encouraged the vigorous growth of grassy weeds. There is a rapid shift of weed flora from dicots to mimetic grassy weeds due to changed micro-environment. Selective effect of hand-weeding considering the difficulty to recognize mimetic grass weeds at an early tillering stage.

The high soil water and nutrient requirement of dwarf wheats compared with that of tall indigenous wheats indirectly gave competitive advantage to N-responsive grassy weeds. The heavy amounts of nitrogenous fertilizers also favoured some nitrophilous dicot weeds such as *Chenopodium album*. The sunlit conditions in the crop stand of the dwarf wheats has had an significant influence on the weed flora. The high crop plant density decreased sunlight intensity to low-statured dicots than grassy weeds which attains more heights than crop plants. This was caused shifting of weed flora from dicots to grassy weeds.

Phalaris minor is a predominant of heavy soils, particularly in rice growing areas, whereas *Avena* spp. infest more in light textured soils.

5.3 Ideal conditions for the heavy infestation of *Phalaris minor*

- *P. minor* does not germinate when temperature is above 27°C. Tall wheats are sown at the end of October and can grow sufficiently before emergence of *P. minor*.
- Late sown HYV wheats are sown in November when mean daily temperature is around 25°C.
- The late sowing of wheat (second half of November) crop provides ideal temperature conditions for its

emergence and the competitions offered by the crop is poor.

- *Avena* sp. thrives better on light soils and less, compacted soils than that of rice-wheat rotation.
- Difficulty of hand weeding due to morphological similarity.
- Shedding of weed seeds 3 weeks before wheat harvest.
- Induced seed dormancy coupled with adoption of a new agricultural production strategy.
- Early first irrigation.

5.4 Weed shifts in maize

Several broad leaf and cyperaceous weeds are the main problem in maize fields. The weed *Rottboellia exaltata* is a native of India, where it is of minor importance in field crops because it is balanced by certain ecological factors and antagonists. The weed usually grows in the medium moisture conditions having a precipitation of 800 mm per annum. This tufted annual grass is becoming problem in irrigated fields crops particularly in tropics.

It is very aggressive in maize and sugar cane fields of Latin America. Removal of the weed by manual weeding is a cumbersome process. The leaf sheath contains very sharp silicous hairs which resembles fibre-glass needles. These needles penetrate through cloths and injure the operators. Sometimes the manual labour suffer with painful infections on hands and legs while weeding. That is why the hired labour refuse to enter the fields of maize if the crop is infested with mature weed stands.

The problem of the weed is started after the introduction of new production technology in maize. In this crop, triazines like simazime and atrazine are commonly used as preemergence herbicide to control weeds. *R. exaltata* though a minor weed but become resistance to triazines. Lack of competition from neighbouring weeds favoured the quick multiplication.

5.5 Prevention of herbicide resistance in weeds

As the use of herbicides increases over the next decade, resistance is expected to become gigantic problem for crop production. The problem can be mitigated by adoption of integrated weed management rather than over reliance on chemical based herbicides. Some of the suggested ways for stopping herbicide resistance are as follows:

- Scientific crop rotation may be adopted with crops of different management practices.
- Diversified chemical control programme may be planned to kill weeds effectively.
- Low and medium herbicide application rates delays evaluation of herbicide resistance.
- Mono-culture of crops of the same field may be avoided.
- Herbicides may be rotated every season to provide different modes of action on the same spectrum of weeds.
- Encouraging the use of herbicide mixtures (tank or proprietary mixtures) for providing broad spectrum of action on weeds.
- Modifying the planting methods, from direct seeding to transplanting in alternate years.
- Controlling of weeds with other agronomic practices like deep ploughing, harrowing, levelling and water management etc.

5.6 Simple herbicides favours the shifts towards purple nutsedge (*Cyperus rotundus*)

Routine use of herbicides for several seasons in a fixed cropping system may cause shifting of weed flora one morphological group to another. Sometimes serious changes may occur among weed communities in an unfavourable direction. In natural arable lands, the usual succession of plant communities is very often from broad leaf weeds to grasses and from annuals to perennial weeds. Continuous use of simple herbicides like 2, 4-D or MCPA kills the annual broad leaf weeds and retain grasses and same perennial sedges. Such type of herbicides hasten the process of weed flora shifting unlike other cultural practices.

Cyperus rotundus is a cosmopolitan weed, spread over most of the topics and a serious problem in a large number of crops. Modern weed control technology contributing to weed shifts and caused increased incidence of *C. rotundus* by eradicating the other stronger competititors in the plant community. In a mixed weed populations, several broad leaved weeds compete strongly with *C. rotundus* and suppress its growth.

5.7 Long range effect

Use of cheap dicot weed killers in the crop is temporarily successful but less attractive on a long range because of increased populations of cyperaceous weeds. The weed shifts from grasses to sedges and dicots to sedges due to use of single herbicide continuously results in long-term failure in weed management. Few herbicides can control purple nutsedge successfully but others remain partly effective because of the following reasons:

- Being a C_4 plant, the plant is efficient utilizer of nutrients and water.
- Multiply through both seeds and rhizomes.
- Having extensive underground rhizomatious network.
- Vegetative propagules can survive in a dormant stage.
- Increased infestation due to dismembering of underground network.

5.8 Herbicidal resistance of weeds

Herbicide resistance is the inherited ability of weed plant to escape kill by a chemical herbicide. Resistance is a relative term and assumed at normal recommended doses of a chemical. Under suboptimal doses, majority of weeds can escape the death but at the same time no weed can survive under heavy rates of herbicide application.

5.9 Evaluation of herbicidal resistance

- Mutations always occur in nature. Chemicals and irradiation plays major role in mutation process. In general non-lethal mutants are at a disadvantage position in a herbicide-free environment.
- Non-lethal mutants are unable to compete with more fit genotypes.
- Resistance plants are selection of naturally occurring biotypes.
- Resistant individuals are very rare but compete successfully within the species.
- The resistant individuals create resistant populations.
- Some chemicals like sulfonylurea may create resistance within a short time.
- Resistance to sulfonylurea may appear within 3-5 years period due to singlesite inhibition (aceto-lactate synthase).
- Auxins, carbamates, chloroacetamides are multisite inhibitors and frequency of resistance to these chemicals are less.

5.10 Process of chemical resistance in weeds

- Repeated use of same herbicide in a particular field may cause a shift in the weed spectrum.
- Susceptible weeds disappear soon.
- Other subspecies and biotypes within a species will fill in the vacancies left by susceptible weeds.
- The resistance weeds multiply and spread rapidly and become major community over a time.

5.11 Factors that favour the speedy development of herbicide resistance weed flora

- Application of highly selective and effective herbicide continuously which kills most of the susceptible biotypes.

- Rapid population shift in favour of resistant biotypes.
- The rate of enrichment of herbicide-resistance population depends upon degree of effective kills.
- Presence of naturally occurring resistant biotypes within the native weed population that enhance under selection pressure.
- Monoculture cropping is an ideal environment for the development of herbicidal resistance.
- Uninterupted use of same herbicide in the same crop over the years.
- Resistance species are at higher rate in temperate climates compared with tropics.
- Intensive use of herbicides that have one specific target site.
- Non-use of herbicidal mixtures in the same cropping pattern.
- Use of higher doses of herbicides increase the selective pressures for resistance.
- The longer a herbicide remains active, the greater its selection pressure.
- Non-use of cultural practices like good land preparation, and water management for the control of weeds.
- Lack of diversified herbicide programme in the crop rotation.
- Over reliance on chemical inputs for the plant protection.
- Non-exploitation of wide range of weed control measures like integrated weed management.

5.12 Resistance weeds to herbicides

The most important result of continual application of one herbicide is the cause of resistance in weed species to chemicals.

Butachlor-resistant barnyard grass (*Echinochloa crus-galli*) was found in rice fields of China where butachlor has been applied for 12 years where two rice crops are grown per year.

The grassy weed, *E. crus-galli* has become resistant to weedicide in rice fields of Greece where propanil (2.0 kg/ha) has been applied for more than a decade. These resistant species was not killed by even at 8 kg ai/ha.

5.13 Ecological fitness

Resistance to propanil has been found in jungle rice *Echinochloa colona* in United States recently. Herbicide resistance biotypes have lower ecological fitness than susceptible biotypes. This differential fitness would allow a susceptible population to replace resistant biotypes in the absence of herbicide. In Japan, propanil-resistant jungle rice showed a trend toward lower reproductive fitness, which may reduce its ecological success. Interestingly, propanil is an important herbicide in direct-seeded rice for the control of barnyard grass since one decade. No propanil resistance is not yet observed in Korea.

Application of propanil (3.4 kg/ha) was started from 60's onwards to control grass weeds in Arkansas rice. After 30 years of continuous treatment, barnyard grass was found to be resistant to propanil. These resistant weed populations were not controlled by application rates 3 times higher than normal. Propanil resistance was noticed more often in continuous rice culture than it was rotated with pulses and soyabean.

Bensulfuron, an effective killer of annual aquatic broadleaf and sedge weeds, was introduced in California rice fields in 1990. The herbicide is widely adopted in all most all of the rice hectarage owing to its effective control. Surprizingly, within 4 years of its introduction, several weed species were confirmed to be resistant to sulfonylureas. The resistant weeds could not be killed even at very high rates i.e. ten times more than normal.

Sulfonylurea-type herbicides like bensulfuron or pyrazosulfuron were also introduced in Asia. Intensive use of these herbicides produced herbicide-resistant ecotypes within few years because they are inhibitors of the ALS-specific target site.

5.14 Resistant biotypes

Reports are also available on the development of weed resistance against rice herbicides like 2, 4-D in Philippines and Malaysia. 2, 4-D-resistant *Sphinoclea zeylanica* and 2, 4-D-tolerant *Fimbristylis miliaceae* were noticed in Philippines and Malaysian rice fields recently.

Existence of sulfonylurea-resistant biotypes of *Sagittaria Montervensis* and *Cyperus difformis* in california rice fields was also confirmed due to lack of crop rotation and by the use of same mode-of-action herbicides.

Alternate cultural practices like tillage, water and nutrient management and weed chemicals with different mechanisms of action are required to control resistant weeds. Management strategies to break weed cycles, such as direct seeding, may provide partial solutions to the problem.

6

WEED MANAGEMENT IN CROPPING SYSTEMS

'What's sauce for the goose is not soup for the gander?'

—Spanish proverb

Weed control in cropping systems needs to be considered on a different bases from that of monocultures. Farmers think that weeds are stable mixtures of familiar that can be controlled easily by conventional methods. However weed growth is not static and difficult to control. Several researchers fail to recognize the tremondously dynamic nature of weed populations. Change in cultural practices of crops results in rapid shift in weed species.

6.1 Composition of weed flora

It is necessary to understand and recognize the shifts in weed populations in various cropping and farming systems. The complete growth cycles between crops-weeds during turn around time will greatly influence the composition of weed flora. This may influence the succeeding weed community. Under intensive cropping, the weed community is the resultant of crop management and crop-weed association in the field.

6.2 Susceptible species may perish

Sometimes various management practices can decrease the heavy populations of problematic grassy and sedge weeds overtime in an intensive cropping system. For example, rotation of lowland and upland crops caused to decline the aquatic weeds like *Monochoria vaginalis, Cyperus difformis, Sphenoclea zeylanica,* and *Ludwigia octovalvis.* Application of butachlor followed by one hand-weeding coupled with high density planting decrease the population of *Echinochloa* spp. Similarly favourable weed shifts have occurred under sole upland conditions. Continuous upland crops in the same field may enhance the populations of *Cynodon dectylon, Cyperus rotundus,* and *Portulaca oleraceae.*

6.3 Shifts from susceptible to resistant species

The pernennial weed *Sonchus arvensis* is resistant to 2, 4-D and dicamba application. However, the annual form *Sochus oleraceus* is susceptible to the afore-mentioned chemicals. Even the inter-specific crosses between perennial *Sonchus* and annual *Sonchus* are more tolerant of 2, 4-D. Differential response of biotypes of *Senecio vulgaris* to simazine and atrazine is also common. The new varieties of *Setaria viridis* with different following durations also shows different response to weed control measures. Hence continued use of particular herbicide can allow a shift in weeds from susceptible to more tolerant species.

6.4 Transport of weed seeds and rootstocks

Man always searches for new crops and new varieties from new continents and new areas. In this search he transports weed seeds and rootstocks unknowingly from one continent to another. A pernicious annual grassy weed, *Rottboellia exaltata* is not a native weed of Bolivia. It infests upland rice, sugarcane, sorghum and maize on thousands of hectares in Bolivia recently. No farmer in the region remembers seeing it more than 10 years ago. It probably was imported from U.S.A. in contaminated grain. In Arabia, several perennial weeds like *Oxalis latifolia, O. corymbosa,* and *Cyperus rotundus* are introduced through imported citrus and mango trees. *Sorghum helpense* was came along with imported grape fruit trees.

6.5 Tackling weeds with weed ecological approach

Because of their highly aggressive characteristics, wide range of adaptation, most perennial weeds in the tropics need more ecological studies to tackle them in an efficient way. In spite of many excellent studies on noxious weed Cyperus rotundus, control of that weed, leaves much to be required. Attention should also pay with *Cynodon dactylon, Imperata cylindrica, Paspalum conjugatum, Sorghum helepensa* and *Scripus maritimus* which are difficult to control in cropped lands. In recent years, the annual grasses including *Eleusine indica, Digitaria sangunalis, setaria* spp. *Rottboellia exaltata, Echinochloa* spp. have become more serious problem in several upland field crops like maize, sorghum and sugarcane. In Bolivia, sugarcane may have as many as 7 or 8 germination flushes of *Rottboellia exaltata* per calendar year. Many weed seeds germinate after the affects of a herbicide have dissipated.

6.6 Multiplication of millions of weeds

The annual broadleaf *Portulaca oleracea* can produce thousands of seeds per plant. In heavily infested fields more than 200 million viable weed seeds per hectare in one year's time. These seeds are often sets seed very early, assuring survival. Seed dormancy is also varies in seed lot.

The parasitic *orobanche* and *striga* cause severe losses in many regions of Asia and Africa. Pitifully little is known of their biology or ecology.

Physiological studies on weed seeds and vegetative propagules for all groups is also needed to know the mystery of their reproduction. Seed dormany, germination pattern, soil longevity, populations, production rhythm and disseminations are important characteristics which needs an urgent study. Students should be placed with emphasis on the above subjects. The on-farm research is highly useful to know the knowledge of traditional weed control practices and cropping systems in relation with weed problem.

6.7 Future strategy

Traditional cropping system throughout the world are the final result of a series of adaptations made by farmers to conditions within their immediate environments. The systems are dynamic and complex. They evolve constantly, adapting to some changes occurred in crops, varieties and management practices. The farmer is an integral part of the information-developing process. His folk wisdom derived from centuries of survival of his predecessors in a harsh environment was considered valuable and can be used in planning and executing the future weed control technology.

6.8 Ominous increase in obnoxious weeds

The change in cropping systems appear to be bringing about speedy changes in weedflora. Currently, there is an alarming increase and spread of perennial sedges in irrigated rice in temperate Korea, Japan and in tropical countries like Indonesia, Philippines and Malaysia. The appearance of perennial sedge *Scirpus maritimus* is a recent problem in Asian rice fields due to rice monocropping and wet condition of field throughout the year.

6.9 Threat of perennials

Other perennials pose threats in tropical world are *Sorghum halepense, Cynodon dactylon* and *Cyperus rotundus* in all upland crops. In plantation crops *Imperate cylindrica, Panicum* spp. and *paspalum* spp. are becoming serious due to continous herbicide use. The perennial *Panicum maximum* ominously increased in sugarcane plantations due to use of chemicals. Many weeds are acquiring tolerance for chemical under intensive herbicide use.

6.10 Strategy in cropping system

Researchers should investigate the following problems in finding the solutions of weed control in various cropping systems:

- What is the weed composition ?
- Are there any changes in weed flora over the years ?
- Why the shifts takes place in the cropping system ?
- What changes in the cropping system have taken place to allow these perennial weeds to become dominant ?
- What changes are necessary to stop the process of shifts ?
- Could the problem have been predicated earlier ? If not, why ?
- What are the requirements of an *"early warning"* system that spots weeds ?

6.11 Threat from annuals

The problematic annual weeds which have been building up in certain crops because, of their tolerance for less expensive herbicides like 2, 4-D, paraquat, and atrazine in developing world. Extensive use of paraquat in banana, citrus and coffee has led to dominance of *Parthenium hysterophorus* in Cuba. *Rottboellia exaltata* has becoming a notorious problem in sugarcane, maize and upland rice in many Latin American and Asian countries. *Phalaris minor* has been enhanced rapidly in 2, 4-D, and isoproturon-treated plots of wheat fields. The weed parthenium usually controlled by spraying paraquat which results in killing of seedlings. However established and growth up stands are resistant to chemical and multiply enormously.

6.12 Ecological concept of weed control

Every weed is a biologically unique entity and should be understood on that basis only each species has its own morphology, ecology, life cycle, reproduction, physiological processes. For sustainable weed management answers are needed for the following questions.

- How, when, how many times should weeding be carried out ?
- Which weeds can be controlled and which cannot ?
- Which weeds can be ignored ?
- What are the susceptible stages of weeds for herbicides?
- What are the resistant weed species for herbicides ?

7

IMPACT OF HERBICIDE USE ON ENVIRONMENT

'Every bullet has its billet' —*French Maxim*

In crop production system, herbicidal chemicals are used to suppress the unwanted effects of weeds (non-crop plants) growing interspersed with crop plants. Herbicides are able to kill the unwanted vegetation selectively and leaves desirable crop plants undamaged by influencing their physiology. The selectivity of chemiclas are mainly due to differences in physical and biochemical characteristics of crop plants and weeds.

Any chemical will damage life at some level of concentration according to primary precept to toxicology. Chemicals that are intended for plants also will be toxic tothe fauna and microflora of the soil. According to second precept of toxicology is that chemical effects will be in proportion to doses. High concentration of chemical may show more adverse effects on living organisms.

7.1 Persistency of chemicals

Concentration of chemical in crop fields will vary broadly, depending upon type of chemical, type of formulation and rate of application. Climatic factors such as temperature, rainfall and humidity are also responsible for the time of persistence. The physico-chemical-biochemical forces also determine the life of chemicals applied to the soil. The dissipation of chemical is slow when the concentration is declined. In fact, the time required for a concentration of three parts per trillion to be reduced to/ part per trillion is similar to three parts per million to be reduced to 1 ppm. However, at any cost, the concentration can never be decreased to zero.

The longer the chemical concentration in the soil or water system, the highter the risk of adverse environmental impact.

7.2 Nature of damage to fish

Irregular and abnormal movements can be seen if herbicides are present in water at sublethal levels. Mercuric, copper and aresenide based herbicides damage the sensory systems and adversely effects the feeding, spawning and hatching functions. These chemicals also destroy the respiratory system also.

Nitrofen is highly poisonous to several species of fish. Massive fish killings are usually happened if rice field water treated with nitrofen flowed into nearby fish tanks. The application of Pentachlorophenol or quick lime to eradicate weedy fish and shrimps may also kill economically important fish. The persistence of herbicides also depends upon pH of soil and water. Nitrofen applied to lowland rice fields of academic nature has a half-life of more than 200 days but it is only one day if the waters are slightly basic in nature.

In general, phenoxy herbicides like 2, 4-D and MCPA are found to be less toxic and triazine group and alachlor are highly toxic to common carp. Herbicides belongs to organophosphates can inhibit the activity of cholinesterase and causes abnormal behaviour of fish in the water.

7.3 Herbicidal toxcity on fish

Pesticide toxicity is mainly depends upon chemical nature and degree of toxicity. In general, herbicides are relatively less toxic to fish when compared with insecticides, germicides and molluscicides. Recently, waste discharges from urban and rural areas and farm lands has resulting in river canals, irrigation streams, natural lakes and farm ponds. The fish living in canals, streams and ponds facing the threat of chemical pollution. Several fresh water lakes and commercial aquaculture areas are slowly deteriorating due to intensive use of herbicides in crops.

7.4 Behaviour of fish against pollutants

Among pesticides, insecticides are highly toxic, molluscicides moderately toxic and herbicides are less toxic to various fishes.

Several species of fish and shell fish have highly developed senses of smell and taste. These creatures usually avoid the polluted waters with the help of their keen senses and goes into clean waters. Clamus may close their shells when they encounter herbicidal contaminated water. Avoidance behaviour to less toxic herbicides helps to escape the hazard of pollutants. However, fish cannot escape the lethal concentrations of chemicals such as phenol and cyanide which results in heavy mortality.

7.5 Sensitiveness of fish species

Sensitiveness to chemicals varies with fish species. Grass carp, silver carp and tilapia are highly sensitive to pentachlorophenol and quickly escape these chemicals but less sensitive to benzene phenol. Crucian carp (*Carassius auratus*) in moderately sensitive to ferrous sulphate and silver carp (*Hypophthalmichthys molitrix*) is highly sensitive to copper sulpahte. The avoidance index is zero for grass carp (*Ctenopharyngodon idella*) and tilapia (*Oreochromis niloticus*) against trial fuilds of benzone.

7.6 Effect of herbicides on azolla growth

Nitrogen fixing cyanobacterium living in azolla leaves carryout its activity through symbiosis. The fern gives room and food material to the algae and in turn algae synthesise the atmospheric-N into mineral form and supply to the host plant. Owing to its great potential in improving the soil fertility of rice fields. Chinese are using this aquatic fern traditionally for fertilizing the rice fields. As it is highly susceptible to the damage of several insects, application of insecticides usually favours azolla growth.

7.7 Detrimental effect

However herbicides more often have a detrimental effect particularly on *Azolla mexicana*. The herbicides belonging to pheolics and bipyridilium are most detrimental to the azolla growth. Even at 0.1 ppm also, they cause about 75 per cent reduction in N_2 fixation and nitrate reduction. Simazine at 10ppm accelerate the nitrate reduction by 20 fold but cause reduction in N_2 fixation by 99 per cent. Benomyl and chloramben at 10 ppm also reduce the N_2 fixation by 91 per cent. The herbicide group belonging to the family triazine, dinitroaniline, urea and benzoic compounds causes reduction in N_2 fixation when applied at concentration of 0.1 to 10 ppm. Surprizingly, Naptalam has no effect on growth and N_2 fixing activity of azolla even at higher concentrations (10ppm). In general, preemergence by herbicides applied about one week, before azolla inoculation has little effects on fern but post-emergence herbicides have a detrimental effect on N_2 fixation. In view of the above, simultaneous applciation of herbicides and azolla inoculation was not a sound practice to get the benefit from both of them.

7.8 Replenishment of carbon and nitrogen through cyanobacteria

Cyanobacteria and microalgae are primary components of photosynthetic aquatic biomass in shallow water ponds, drainage ditches and wetland rice fields. In deep water bodies like lakes and reservoirs, macrophytic algae and vascular macrophytes are important components of photosynthetic production of carbon compounds. The mean primary photosynthetic production usually ranging from 0.5 to 1 g cM^{-1} d^{-1}. Microalgae trap significant quantities atmospheric carbon and nitrogen and incorporate in the soil. This type of fixing and incorporation organic matter and nitrogen helps in enhancing the soil fertility and productivity.

7.9 Functions of blue-green algae in wetland rice fields

- Radiation dependent biological nitrogen fixation by blue-green algae contributes 5-20 kg N ha^{-1} per crop season from free living forms and 30-60 kg N ha^{-1} per crop from symbiotic forms associated with azolla.
- Nitrogen is temporarily immobilized and recycled through grazing and death. The immobile N is available to higher plants after decomposition.
- Replenishment of soil microbial biomass.
- About 30 per cent of N contained in algae and 20 per cent of aquatic plants that decomposed at the surface soil will be provided to the rice plant.
- Microalgae deplete CO_2 in flood water and increases pH values of flood water which induce ammonia volatilization (2 – 60 % N)

7.10 Influence of herbicides on cyanobacteria

Application of herbicides, in general, inhibit indigenous or inoculated biological-N fixing microorganisms. Scanty information is available on the interaction between pesticides and legal inoculation.

Pentachlorophenol (PCP), a pesticide used in rice, acts as an insecticide and herbicide as well. The chemical has inhibitory effect on diatoms and blue-green algae when applied on surface soil. Incorporation of PCP in the soil has no effect on the growth of algae. Algae growth of rice fields will be declained due to application of C N P (2, 4, 6 - trichlorophenyl 4 - nitrophenyl ether). Few herbicides like butachlor adversely affect N-fixing ability of blue green algae. Numerous herbicides have a detrimental effect on photodependent biological N-fixing organisms. However most of the algae species can tolerate up to 100-500 ppm of 2, 4-D. The dose is several times higher than that recommeded for rice fields.

7.11 Negative effect on algae

Application of butachlor @ 5-30 kg ha^{-1} did not show any significant impact in algae inoculated rice fields. When algae inoculation was effective, most of the herbicides had not effect on soil nitrogen fixation. Negative effect on algae multiplication can be seen when propanil was sprayed in rice fields. Recommended levels of Fluchloralin, thiobencarb and TOK had no effect on the multiplication of blue-green algae. However 2, 4-D sodium salt, 2. 4-D amine and propanil decrease the algae blooms to an extent of 40 per cent.

7.12 Role of herbicides in aquatic food chains

Herbicidal chemicals indirectly show their impact on food chain of aquatic habitats. The fish living in natural water bodies depends upon natural flora and fauna. The food supply of fish usually influenced by feed and manure applied to waterbodies. Phytoplankton usually grown in still waterbodies is mainly eaten by filter-feeding fish. Herbivorous fish feed primarily on aquatic flora. Benthos are eaten by omnivorous fish. All fry and fingerlings of fish feed on Zooplankton in their juvenile stages. When herbicides are applied to the pond, these chemicals show their direct or indirect effects on the food chain links. Some food chains may be disrupted which in turn indirectly influence the fish growth. For example when algicides are applied to the pond for the control of excess green algae films, all phytoplankton will be destroyed. Filter-feeding fish which depends upon these algae may be starved due to lack of food.

7.13 Effect on phytoplankton

Sometimes atrazine and simazine are sprayed to kill the excess growth of coontails (*Ceratophgllum* spp.) pond weed (*Potamogeton* spp.) and Najas (*Naiad najas*) causes the temporary decline in phytoplankton. After the decaying of aquatic plants, the water blooms will reappear profusely. Excess growth of phytoplankton decrease the oxygen supplies during sunless days and cause fish mortality.

Killing of water weeds through herbicides adversely affects the population of some aquatic animals which feed on them. For example, treatement of ponds with diquat (100 ppm) to control *Elodea* weed adversely influence the population of molluscs. Because most of the snails and gastropods feed on Elodea plants will suffer due to food shortage. Similarly Nymphs of dragon fly, damsfly and Ephemeropteran and larvae of stonefly disappear shortly when monuron was applied to the ponds because of the eradication of their food source. Surprizingly, high densities of oligocheate worms and chronomus was common when fenoprop was spreyed on fish ponds. Sometimes these worms will be enhanced up to 10-fold after chemical treatment.

7.14 Killing of pond weeds

Application of certain herbicides may change the composition of aquatic flora population. When dichlobenil was sprayed to control water weeds, all pond weeds will be killed but *Chara* may recover and flourish, filling the niche of other pond weeds. Application of diquat for killing the *Elodea* may enhance the population *nitella.* Owing to change in aquatic flora, the herbivorous fish which have selective feeding habit may suffer with starvation.

The heavy metals that are present in chemical compounds may accumulate in large quantities in aquatic flora and fauna in regularly treated ponds. Alligator weed (*Alternanthera phyllcxeroides)* can accumulate copper ions to a concentration of 30,000 times more than the concentration of water if ponds are treated with copper sulphate. Similarly water meal (*Wolffia* spp.) accumlate cadmium and lead ions to a concentration of 4800 and 2000 times greater when they grow in chemical treated water.

7.15 Biodiversity of paddy fields

Application of herbicides usually decrease the biodiversity of organisms present in paddy fields due to disappearance of permanent reservoirs of organisms in the vicinity of the fields. Surveys indicated that about 183 species of aquatic invertibrates were recorded in traditional rice fields of Thailand, Malaysia, Philippines and India. The population dominance of these organisms are inversely proportional to diversity. Among the invertibrate community ostracods, chironomids and molluscs are usually dominate when herbicides are used. The marked decrease in number of taxa in rice fields is an indication of decline in species richness. It is usually accepted that crop intensification and use of herbicides decrease biodiversity and provoke "blooms" of certain organisms.

7.16 Change in flood water fauna

Addition of thiobencarb drastically reduce populations of cladocerans, odonatans, midges and mosquito larvae. Resurgence of cladocerons mosquito larvae is also common after sometime. Butachlor and carbofuron do not affect the flood water invertibrates significantly. Conventional rice pesticides, in general, not affects the snail acitivity of rice fields but their population sometimes increases due to reduced competition.

Saprophytic and parasitic nematodes will not affect due to application of thiobencarb. Disappearance of aquatic oligochaetes is thought be the result of PCP (Pentachlorophenol) instead of NIP (2, 4-dichlorophenyl P-nitrophenylether). Reappearance of oligochaetes takes place when CNP (4-nitropheny 1 2, 4, 6-trichlorophenyl ether) and thiobencarb is used.

7.17. Shifts in algal strains due to herbicides

Sensitivity to any herbicide may vary considerably among algal strains. Herbicide application causes shifts in dominant strains within the cyanobacterial community rather than decline in the algal biomass. Some micro-algae and cyanobacteria show resistance to pesticides because of their adaptation to increased concentrations. Spontaneous mutants are usually resistant to herbicides.

7.18 Effect on N-fixing organisms

Usually herbicides can inhibit the cyanobacteria and photodependent-N fixing organisms. The herbicide Pentachlorophenol (PCP) is inhibitory to cyanobacteria when applied on the surface soil. Several chemicals like CNP (2, 4, 6–trichlorophenyl 4-nitreophenol ether) inhibits photodependent N_2-fixers and algae growth. Sometimes butachlor affects the N_2 -fixing ability of N_2- fixing strains. Several species of algae can tolerate up to 100-500 ppm of 2, 4 -D which is much higher than recommeded rate. When algae inoculation was effective, herbicide application had no effect on soil- N. Recommended levels of thiobencarb, fluchloralin, TOK had no effect on cynobacteria inoculation. 2, 4-D Na, Propanil, weedone (2, 4-D) application sometimes decrease the algal activity by 15-40 per cent.

Certain herbicides like S-triazines and substituted ureas alter phytoplankton composition by selecting inhibition of certain species.

7.19 Activity of aquatic invertibrates

Flooded rice field is usally colonized by invertibrates like *rotifers, molluscs, oligochaetes, ostracods, copepods, cladocerans,* fish fingerlings, shrimp and nematodes. They play significant role in nutrient exchange from soil to water medium, nutrient recyclers, pest predators, pests and vectors of human and animal diseases.

Oligochaetes, particularly organisms belongs to *tubificidae,* which is primary component of *zoobenthos* helps in the exchange of nutrients from soil and flood water. This results in augmentation of nitrogen absorption by rice plants. On an average rice fields have a population of about 40,000 m^{-2}.

After transplanting, there is a proliferation of phytoplankton in first fortnight. Thereafter microcrustaceans and mosquito larvae appears and consume the blooms of microalgae within two week period. These primary consumers mostly graze on photodepedent N-fixation algal species whcih indirectly helps in dominance of noncolonial forms. N-fixation is very less in non-colonizers. Ostracodes usually recycle the N from flood water to an extent of 20 kg N ha^{-1}.

7.20 Detrimental effect

Some of the aquatic invertibrates have harmful effect in rice ecosystems. Mosquitoes occurred in marshy environment are vectors of diseases such as encephalitis, malaria and filaria, snails like *Bilinus* spp and *Liminease* spp. are vectors of bilharziosis. Chironomids feed on rice seedling roots and damage them. Herbicide application alter the nutrient cycles.

8

HERBICIDE PERSISTENCE AND HUMAN HEALTH

He who does not take time by the forelock, will soon find suffer.
—Latin proverb.

In crop production, herbicides are used to kill the noncrop plants growing interspersed with crop plants to avoid their undersirable effects. The herbicides are able to damage the weeds and leaving the desirable plants under undamaged due to differences in physical and biochemical characteristics of plant species. Most of the chemicals act on the photosynthesis and respiration system of the weeds.

8.1 Precept of herbicide technology

A basic precept of herbicidal technology is that some level of any chemical will damage the plant life. Though the herbicides are designed for killing the weeds but they also toxic to native fauna and microbes. A second precept of herbicides is that effects will be in proportion to rate of chemical. Higher the concentration of chemical more will be the adverse effect on flora and fauna of the applied field.

8.2 Principles of herbicide dose and its persistence

- Concentration of herbicides in crop fields may vary due to nature of chemical, rate of application and type of formulation.

- The dosage of chemical also depends upon climate and soil type.
- Herbicides dissipate over time, in response to physico-chemico-biochemical reactions.
- Dissipation of chemical is slow as its concentration decreases.
- The more persistent a herbicide, the greater the risk of adverse environment.
- Concentration of herbicide will never be reduced to zero under any circumstances.
- Toxic impact of herbicides are not instantaneous.
- Toxicity of herbicides can also be seen on non-target organisms which depends on nature of chemical.
- Herbicides dissolve in water and undergo dilution.
- Turbidity of water reduce the chemical activity of herbicide.

8.3 Damage of fish by drainage flows

The degree of toxicity on fish due to herbicides mostly depends upon nature of chemical and its concentration. Farmers usually apply herbicides for the control of various types of weeds. The drainage water from rice fields flowed into nearby fish ponds, canals and lakes and causes slightly to severe damage to the aquatic fauna. Some herbicides may cause massive fish kill. In general insecticides are most toxic to fish, molluscicides are more toxic and herbicides the least toxic. Some herbicides can be used as germicide or as insecticide and molluscicides based upon the chemical nature.

8.4 Pollution of streams

Application of herbicides in crops grown on watershed area of rivers results in pollution of streams and rivers. The fish present in these rivers and brookes will be affected. Several natural lakes like Taihu lake in China deteriorated in this way and become unfit for fish growing. The persistence of several herbicides in soil and water is rather short. In irrigated rice field, the half-life of molinate and nitrofen is 10 d and 200 d respectively. If the irrigation water is slightly alkaline, the half-life of above herbicides are declined several folds.

8.5 Sensitivity of fishes

Shelfish and fish have high degree of senses of taste and smell. Fishes usually take avoidance and clams may close their shells when they encounter water containing pollutants. The action of avoidance varies with chemical and its doses. Sometimes fish may not avoid some toxic chemicals like phenols and cyanamide which results in mass mortality.

8.6 Nature of damage

The silver carp, grass carp and Tilapia were sensitive to pentachlorophenol but slightly snesitive to benzene phenols. (Table 8.1) Fish confined over a long time in water bodies obtaining slight concentrations of pollutants exhibit abnormal movement or restlessness. Herbicides belongs to organophosphate group may inhibit cholinesterase activity and copper sulphate and mercuric compounds may damage the sensory systems. Arsenide group destroy respiratory functions of fish. All these pollutants adversely affects the feeding, spawning and hatching activities of the most fish species.

Table 8.1. Avoidance of fish to herbicides

Type of fish and its avoidance index	*Herbicides*			
	Pentachloro-phenol	*Benzene*	*Ferrous sulphate*	*Copper sulphate*
Grass Carp	Quickly avoid	Slightly avoid	Not sensitive	Not sensitive
Silver Carp	Quickly avoid	Not sensitive	Not sensitive	Moderate sensitive
Crucian carp	Not sensitive	Not sensitive	Moderate sensitive	Not sensitive
Tilapia	Quickly avoid	Slightly avoid	Not sensitive	Not sensitive

8.7 Contamination of pond sediments

Application of herbicides to control aquatic weeds in static waterbodies results in disappearance of most of the aquatic fauna due to contamination of sediments. When 2, 4-D is used in lakes to control water weeds like *Myriophyllum spicatum* and *Nitella,* the concentration of chemical may reach 100 times more in the sediments due to slow settlement at the bottom after a few days. With the result, the bottom layer fauna like sankehead (*Oppihiocephalus argus),* shrimp (*Macrobrachium* spp.*)* and soft-shelled turtle (*Triony sinensis)* may die due to exposure of high chemical concentrations. Several aquatic organisms living in bottom layers disappear very soon due to contamination of herbicides and chemical factory wastes.

8.8 Ammonia poisoning of ponds

Accumulation of excess ammonia in pond water is harmful to common fishes. Excess ammonia indicates the deficit of oxygen in the waterbody. Mortality of fish under this environment is in two ways. Unionized ammonia is highly poisonous to fish. Even low levels of ammonia (1 mg/litre) may kill carp fishes within few years. The ammonia in the water decrease the ability of blood hemoglobin to combine with oxygen and fish may die due to lack of life giving oxygen.

Application of herbicides like methylurea, dimethylurea, phenylurea belongs to urea group produce considerable quantities of ammonia after hydrolysis. The unionized ammonia fraction enhances with higher pH and temperatures. Fish mortality is usually high in summers owing to unwanted accumulation of ammonia in fish ponds. The ammonia level should not reach more than 0.04 mg/litre in the fish ponds for the safety of aquatic life.

8.9 pH values influence the fish diseases

The chemicals applied to clear water in the ponds indirectly inflence pH values of the water. Fish mortalities occurs usually at low pH (> 2.5) and at high PH values (< 10). Caustic lime (calcium oxide) is very often used by farmers to kill the unwanted organisms in the pond and to clear muddy water. Over application of lime increase the pH of pond water to more than 11 which kills several types of fish. Even hard fish inevitably die at extreme pH values. The pH of pond water also influence the life cycles of several fish parasities. Some parasites like *Costria necatrix, Chilodonella* and *Odinium acidophilum* usually found in acidic waters. Application of copper sulphate to kill algae in fish ponds acidify the water and encourage the fish parasites.

8.10 Herbicide use and health

The use of herbicides in field crop production has been increasing in Asian countries during the past decade. This is mainly due to scarcity of agricultural labour in rural areas. The rural wages are rising rapidly. In some areas, there are shifts from labour-intensive transplanted rice systmes to labour-conserving direct-seeded rice systems. Eventhough herbicide use is ostensibly more profitable in labour scarcity areas, the farm labour may incur significant health hazards. Prolonged exposure to pesticides results in adverse health effects and impose significant costs and reduce labour productivity.

8.11 Herbicide users

Most of the Asian farmers apply herbicide once per crop. Few progressive farmers make a second application to control excessive weed growth. The users exposed themselves to herbicide hazard during handling, mixing and spraying operations. Farmers usually did not use any protective clothing, gloves, mask or rubber, boots when spraying the liquids or broadcasting the granular forms.

8.12 Intensity of exposure

The risk of pesticide exposure mainly depends upon intensive of exposure. Chemical exposure may vary from the intensional poisoning (ingesting large dose at once) to occasional, low level ingestion of chemical residues. Residues enter into the human system through food or drink or breathing.

Unintentional exposure is also dangerous on a long run. Occupational exposure to herbicides over a long time is also quite common. Farmers usually mix the herbicides in bucket with bare hands and spray the chemical half nackedly without any protection clothing. These are highly risky to cultivators. Working in the vicinity of newly sprayed fields, proximity to spray equipment, herbicide containers and contaminated clothing are subjected to risk to some extent. Logit regression analysis indicates that there is a discernible adverse effect on gastrointestinal, neurological and dermal system of human when exposed to chemicals.

8.13 Effect on dermal system

Contrary to common belief, pesticides primarily not enter to the respiratory tract but enter the body through skin easily. Farm workers subjected to dermal exposures while mixing and transferring herbicide concentrates during application. The dermal exposure levels are higher than inhalation levels during spraying operations. Dusting of pesticides leaves residues on exposed skin. The residue amounts vary with application technique, duration of exposure, working conditions and protective equipment. Spraying with spinning dise applicator or an electrodyne sprayer reduce dermal contamination than Knapsack sprayer. Among the limbs, the hands, forearms and legs are mostly exposed to chemical.

Herbicides such as 2, 4-D, acetamides and organochlorine insecticides are mild skin irritants and high sensitizers. The indicator of pesticide exposure is the appearance of exzema, chronic allergic dermatitis. Sometimes, the skin develops lichenification and fissuring. Thickening of skin with accentuated markings is also another health indicator. Loss of distal portions of the toenails which gives them an *"eaten-up"* appearance.

8.14 Eyes are vulnerable

The eyes are sensitive organs and extremely vulnerable to chemical hazard during pesticide application. The commonly used herbicides like acetamides and 2, 4-D are moderate eye irritants. Incidence of eye abnormalities are common to users particularly older people and children. A chronically irritated eye can form a pterygium, a vascular membrane over the cornea. When people exposed to wind and dust, the vascular membrane may encroach on the pupil and diminish visual acuity. Surgery is needed to improve eyesight. The diminished vision would reduce the productivity of farmer to a great extent.

8.15 Gastrointestinal tract

Nobody swallow the pesticide intentionally but they enter the intestine accidentally. Farmers usually wipes off sweat near the mouth may unknowingly ingest traces of pesticide while pesticide application. The herbicides like 2, 4 - D and organochlorines are moderate irritants of gastrointestinal lining, causes vomiting, mouth ulcers and diarrhoea. Carbamate insecticides may cause severe intestinal irritation. Organophosphates and copper salts also irritate the digestive tract, causes vomitting, intense nausea and diarrhoea. Epigastric tenderness and pain also associated with nausea in chronic gastritis. Farmers who receives low nutritional diet are highly susceptible to these irritations.

8.16 Neurological disturbances

Herbicides belonging to phenoxy acids and organophosphate compounds are considered to be neurotoxicants and causes neurological disorders when exposed long time. They are causative agents for polyneuropathy, typically manifested as motor weakness in the distal muscles and sensory deficit. Deep tendon reflexes are lost in the muscle cells. Isolated hyporeflexia and diabetes are also possible when exposed to organophosphorus pesticides. Farmers in old age and who drink alochol are highly susceptible to the malady.

8.17 Respiratory disorders

The incidence of respiratory abnormalities among the farmers are common when exposed to chemial irritants. Incipient lung disorders like cough, cold, sputum formation, wheezing, rales, tenderness and decreased chest expansion are some of the disorders indicating the pesticide exposure. Bronchial asthma and abnormal lung movements indicates the disorder. Persons having smoking habit, poor nutritional status may face severe lung disorders. Farmers believe a cloth face mask provides protection but actually it creates more problems. The cloth absorbs the chemical which the wearer then inhales.

The health costs associated with pesticide exposure overwhelm any productivity gains. The value of the yield gain with pesticide use is several times lower than the medical expenditure for the health impairment.

9

SOCIAL AND ECOLOGICAL IMPLICATIONS

"Every cock crows on its dunghill'

—English Proverb

There is a common agreement among scientists that the labour input for weed control is tremondous. Sometimes it is more than 50 per cent of labour input in several tropical countries. Heavy rainfall and high temperatures encourage prolific weed vegetation which requires high labour input for manual weeding. A shift from broadcasting to row cultivation in some Asia and African countries facilitates using bullock drawn implements. In developed countries, mechanical inter-row cultivation increases the speed of weeding with tractor drawn implements. Recently chemical weeding is spreading in third world countries. More cautions has to be taken before introducing herbicides. Much to be know about long-term effects of herbicides on fauna, flora, man, soil and waters.

9.1 The strain and drudgery of manual weeding

In all developing countries, hand weeding is carried out by members of the farm family itself, whereas on larger holdings hired labour is engaged rather than family members. The strain and drudgery to the many crores of rural poor who are compelled to do this hardwork is mainly for their survival. No alternative employment is available in rural areas. Youngmen does such a boring and straining work as a major part of their life unless circumstances forces them to it. If the herbicides are advocated in the peasant agriculture, there is a risk of loosing even this means of support, when no alternative employment is available. More energy is expended for the weeding of man's crops than for any other single human task.

9.2 Life style of farmer

Throughout the hot tropics, hand weeding is the life style of poor farmers that do not have any other vocational choice. Uneducated men and women throughout the developing nations consider hand weeding as a fall-back employment position. It is very easy for people who are not intimately connected with peasant farming to assume that people who weed on farms for living do it out of choice, but the fact is that a majority of these people weed because the alternative is starvation. As long as large groups of agricultural labourers is present, hand weeding is not a drudgery but means of feeding. Several farmers enjoy the weeding but do not feel drudgery. Starvation is an alternative of liberation of labour from hand weeding. That is why government restricts the import of herbicides in third world countries because of large scale unemployment. In developed countries the life style of hand weeding is considered as both degrading and dehumanizing. In view of net income of crops, chemical weeding does not outweigh the benefits of hand weeding. Sometimes, the drudgery may outweigh possible discomforts associated with handling herbicides. Partial or incomplete knowledge of herbicides creates potential danger for both crops and humans.

9.3 Back-breaking exercise

Weeding of field crops with hand pulling is very straining. In various cropping systems, hand pulling without any mechanical instrument is a stooping, bending, back-breaking exercise that traditional farmers in the tropics have learned to live with. However, children strive to escape this exercise very often. In row crops intercultivation can be done with bullock drawn ploughs which take care of weeds. But row cropping is less common in semi-subsistence economy of the humid tropics. In some of the African and Asian countries, the women often carry the entire crop husbandry work including weeding. Hand weeding in rice with women is very common in India. Some operations like transplanting and hand weeding in rice was done soley by women. Young boys and girls are able to avoid boring weeding work to some extent. Children cannot do the weeding work because they go to school. Finally it is very difficult to avoid for housewives with a feeling of responsibility to feed their family.

9.4 Agro-biological manipulations

In these situations, improved cultural practices with small instruments supported by suitable agro-biological manipulations of important components of cropping system could lead to less straining weed control. Row corpping, improved handtools, animal drawn instruments and cover crops can be used for efficient weed control. There is no need of spraying herbicide or even most important way to avoid unpleasant hand weeding work of farmers or to fill the void of child labour when they go to schools.

9.5 Labour input for hand weeding

Weed control is one of the most labour-intensive aspects of tropical agriculture but it is chemical-intensive in temperate crop husbandry. A large part of total labour input goes to manual weeding in humid and semi-arid tropics. Subsistence farming in the tropics centres around the manual weeding, over 50 per cent of of the farmer's time goes to manual weeding. This leaves him with little time for his recreation, creative thinking and settling his family affairs. While traveling across the tropical world, one may have the impression that one-half of the world's men and women are in the fields. They move slowly, stooped and doing weeding silently. These uneducated and faceless people, symbolize the great mass of humanity. They spend half of their life time in weeding the crops. Many authorities that dealt with weed control in developing world, estimated that manual weeding occupies about half of the labour input required for crop production.

9.6 Weeding work in arid western India

The western Rajasthan receives only a few hundred millimetres of rainfall per year, that to between June and August. The soil has course textured and low moisture retentive ability. In spite of low and erratic rainfall, water accumulates in such amounts in lime stone cavities below the calciorthid soil that it can be pumped from an underground level of about 50 metres. Previously the area is sparsely populated but after the introduction of irrigated farming most of the farmers settled there. Some of the wealthy landowners has invested huge amounts in digging deep open wells and electric pumping sets and tubewells. Weed problem is also enhanced due to moistering of soil. Employment opportunities are created due to expansion of irrigated agriculture. In these areas the land owner negotiate with male contractor about carrying some tasks like manual weeding. The contractor employs a gang of labourers, often relatives and neighbours.

9.7 Organization of labour crews

In peninsular India, the male and female leaders organise large crews of females, during peak periods on daily wages for weeding. The labour leaders negotiate the wages with landlords and send required number to farmer's field. The organiser gets wages equivalent to twice or thrice of female labour for providing labour to the farmer. In several southern states gang system exists. The gangs are led by foremen who negotiate with the cultivators about wage rate and type of operations. Small guys also joins in gangs due to attractive income. Even small farmers also requires hired labour during seasonal labour peaks. In labour shortage areas, farmers fail to weed their crops enough to avoiding weed damage to crop yield.

9.8 Male wages are more than female

Although both men and women participate in hand weeding, farmers prefer women labour for weeding because of their low wages. The casual agricultural operations for the male is somewhat different with the female. Usually men carryout bullock operated intercultivation and turning and soil with crowbars and spades. Some makes have a semi-permanent status of employment in the farm. The young girls are engaged to remove the rhizomes, roots, and shoots of weed from the dug soil with very low wages.

9.9 Hoeing

In dryland crops, labourers use hoeing instruments locally called as *khurphi, kasi* and *tavugala* which is used to hit the lower root zone weeds. The work is intensive and straining in case of perennial roots. The energy spent in this task is not proportional to the benefit obtained through crop yields. Inadequate and inefficient tools are responsible for this situation. Use of herbicides in this situation is not advisable because of farmer's ignorance. Spraying with herbicides needs technical know-how which leads to clumsiness of available equipment. Bullock intercultivation is advisable if the crops are sown in rows.

In heavy soils, hoeing is very difficult, back-breaking, exhausting and tedious because of hard soil. Vigorous hoeing can be done only by strong male gangs. Land-owners usually employ very young girls, in early teens, for hoeing because of cheap wages. These girls get very little money in this work but as a team they cannot avoid.

9.10 Animal hoeing

Traditionally intercultivation by bullocks and he-buffaloes was done in dry crops like maize, pearl millet, pigeon pea, sugarcane and groundnut. It not only destroy the weed growth but also improve the soil conditions. Soil moisture is also conserved. Sometimes, hoeing with *desi* plough has an initial negative effect because of breaking at the internodes, uprooting of seedlings and damage of crop owing to trampling of animals. However, it induces profuse tillering and vigorous growth after irrigating the field. Late intercultivation is not advisable as it reduces crop yields. In Rajasthan and Haryana, camel drawn iron ploughs are used for breaking the soil.

9.11 Increased threatening of perennials

Short-lived annual weeds usually reproduce through seed. Some plants extend their life through vegetative propagation. Few plants carry out propagation through both sexual and asexual reproduction process. Vegetative propagation usually seen in advanced stages of plant succession in nature. Unscientific weed management results in heavy infestations of perennials at the cost of the annual weeds. Minimum tillage with light country ploughs and chemicals encourage the build-up of perennial rhizomatous grasses.

Conventional agricultural equipment is suitable for the control of shallow rooted soft annual weeds but unable to eradicate stoloniferous and rhizomatous weed species. The problematic perennials such as *Cynodon dactylon, Imperata cylindrica* and *Saccharum spontaneum* usually proliferates when the tillage equipment fails to plough the land deeply and unable to break the soil. The problematic weed *Cynodon dactylon* multiply at enormous rates in sorghum fields of heavy vertisols due to use of light wooden ploughs. Tractor drawn inverted steel ploughs are needed to uproot these weeds.

The deep rooted mimetic weed *Saccharam spontaneum* is most troublesome in sugarcane fields. It also take over the wastelands where the soils are not disturbed. The fire-resistant perennial weed *Imperata cylidrica* is becoming highly dominent in grasslands, crop fields and in shifting cultivation where burning is practised. The appearance of *I. cylindrica* is the symptom of land degradation.

However, the cogon grass helps to prevent land sliding of mountain soils by binding soil effectively.

9.12 Improving competitiveness through breeding

Greater competitiveness of crop over weeds is an important component of weed management strategy. Plant breeders should introduce some of the competitive traits while evolving high yielding crop varieties. For example, some of the *characters of competitiveness* in direct-seeded rice are identified as follows :

- Early seedling emergence.
- Greater seedling vigour.
- High and faster growth rate of crop.
- The seeds should have tolerance for anaerobic conditions.
- The pregerminated seeds should capable of emergence in standing water.
- Increased plant height of crop.
- Greater root volume and higher root biomass.
- Longer growth duration.
- Vigorous tillering activity.
- Tolerance for parasitic weeds.
- Should have allelopathic activity that restricts the growth of dominant weeds.
- Should possess herbicide resistance.

9.13 Contradictory characters

However, some of the traits like plant height, longer growth duration and heavy tillering are contradictory with the plant type of "*Super rice*" which is designed for high yield potential. When imparting weed competitiveness in crop, some of the desirable characteristics should sacrificed.

9.14 Traditional tall cultivars–winners in competition

All traditional rice cultivars and land races are good competitors with native weed flora. These cultivars possess certain characteristics like tall stature, droopy leaves, fast growing, high tillering capacity and rapid nutrient absorption which believed to contribute to the weed suppressing power. These rices can successfully compete for resources with weeds. Contrary to this, the present dwarf and semidwarf rices are having the traits like short stature, errect and narrow leaves, and responsive to applied nitrogen which favours high yields but proved to be weak rivals compared with weeds. Even among the modern cultivars, intermediate statured varieties are more competitors than semidwarfs.

The tall, robust and fast-growing grassy weeds like *Echinocholoa glabrescens* in wetland rice and *Rottboellia cochinchinensis* in dryland rice are outgrown over existing high-yielding cultivars recently.

9.15 Head start over weeds

Recently farmers of Asian countries are slowly shifting from transplanting to direct-seeded rice culture due to sacricity of labour. Slow growing dwarf cultivars which do not have the initial vigour cannot tolerate weed competition. Cultivars having high head start over weeds are highly desirable. Varieties which grow rapidly within the first 30 days of the life cycle should be identified for competing with germinating weed seedlings.

9.16 Causes of high competitive ability of weeds

- Higher photosynthetic ability.
- Greater water use efficiency.
- More nitrogen use efficiency.
- Large potential to acclimatize with change in ecosystem.

- Most of the grassy and dicot weeds are belongs to C_4 plant group (Hatch and slack pathway) which are efficient carbon assimilators in tropical conditions.
- Presence of chlorophyllous first leaf in barnyard grass (*Echinochloa* spp.) which gives a 4-day head start in direct-seeded rice.
- Contrary to this, rice seedling has non-chlorophyllous first leaf which cannot assimilate photosynthetes at the early stage.
- High vulnerability and sensitiveness of rice to inter-specific competition in the field ecosystem.
- Weeds are drought resistant and has greater potential to access water under adverse conditions also.
- Suppression of crop plants through allelopathic activity.
- Suitable morphophysiological trait to explore the resources under scanty.

10

INTEGRATED WEED MANAGEMENT

"The principle of integration is the basic law of nature.'

—Charles Darwin.

Integrated wed management system is basically an integration of effective, dependable and workable weed management practices that can be used economically by the producers as a part of sound farm management system. This approach takes into account increasing agricultural production, reduce economical losses, minimum risk to human health and less potential damage to flora and fauna, besides improving the safety and quality of the environment. Integrated weed management is (IWM) not meant for replacing selective, safe and efficient herbicides but is a strategy to encourage a judicious use of herbicides along with other safe, effective and eco-friendly control measures. IWM may act as catalyst in stabilizing the cost and efficiency of agricultural production against inflationary pressures that are created by the increased cost of weed control, failure of weed control practices and increased regulatory constraints.

10.1 Weed management and control

Weed control methods are often used as euphemism for weed management practices but two are not the same. Weed management implies the directing of weed situation by a judicious use of different methods of control, with the intention of decreasing the harm caused by weeds to a level that is economically acceptable. Such a type of management means to reduce the weed population to a level where their presence has no effect on the areas of economic use. Management is the shifting of the crop-weed balance so that yield is not economically reduced. It has also to be considered on the basis of its *economic, ecological,* and *sociological consequences*. The choice of weed control methods not only depends on technical know how but also depends on geographical and sociological factors.

10.2 Integrated approach of weed management

Hitherto using herbicides for the control of weeds in every crop is becoming common. No substitute methods are searching for chemical weed control because of its less labour requirement. Majority of weed scientists determine the efficacy of half a dozen herbicides in crops by comparing with traditional weed control practice or with two hand weedings. Thus weed control trails dealing with chemicals without considering the other cultural methods. Recently the situation of controlling weeds has seriously deteriorated due to continuous application of chemicals. There is wide spread emergence of weeds that are resistant to all currently used herbicides. About 50 undesirable weed species have been found to be resistant to triazine group of herbicides. Prolonged application of chemicals, routinly repeated cultivation practices and practices like minimum tillage has led to change and modifications within the weed communities present in the same field. Increasingly emphasis on integrating various control methods is growing due to social and environmental problems. In this approach, farmers use low cost chemicals in combination with preventive and cultural weed control methods which are *agronomically effective, economically viable* and *ecologically safe*.

10.3 Preventive methods

Prevention is always better than cure. The measures to deny the entry of weed in new areas encompasses preventive weed control.

10.3.1 Sowing weed free seed

Most of the weeds are introduced into agricultural lands through contaminated crop seeds, farm implements, and transported soil. Several weed species have a life cycle similar to crop cycle and produce enormous seed at the crop harvest. The weed seed mixed in crop seed and enters into new areas wherever it is sown. Even small weed seed quantities multiplies several fold when sown in crop fields and become a serious problem. It is estimated that even one per cent weed seed in cereals will become 1 kg weed seed/hectare and spread in vast areas through crop sowings.

Weed seed dispersel through crop seed can be arrested through

- Weeding in crop fields particularly before harvest.
- Weeding the fields before flowering of weeds.
- Cleaning of crop seeds with sieving and blowing during processing.
- Prevention of soil transportation from weed infested fields during levelling.
- Harvesting and threashing implements should be cleaned thoroughly.
- Adopt all techniques to arrest the entry of weed seed through animals, manures and other agencies.

10.4 Objectionable weed seed

Some weed species are designated as objectionable according to Seed Act, 1966 and certified seed should contain not more than 0.01 to 0.1 per cent depends upon weed species.

Crop	Objectionable weed	Permissable limit of Weed Seed (%)
Brassica compestris	*Argemone mexicana*	0.1
Oryza sativa (Rice)	*Oryza sativa* var. *fatua*	0.01
Trifolium alexandrinum (Barseem)	*Chichorium intybus*	0.05
Trigonella foenumgraceum	*Melilotus alba*	0.1
Triticum aestivum (Wheat)	*Convolvulus arvensis*	0.01
Medicage sativa	*Cuscuta* spp.	0.05
Brassica juncea (Mustard)	*Argemone mexicana*	0.01

10.5 Use of certified seed

Supply of genetically pure seed of crops to the farmers helps in prevention of weed seed contamination. Seed supplying agencies of either public or private sector should take adequate measures to produce pure certified seed.

10.6 Legal measures through quarantine laws

Legal procedures are necessary to arrest inter-state and inter-country movement of noxious weeds. Inadequate quarantine laws and lack of initiation to implement them caused the rapid spread of several foreign weeds. The weeds such as *Poa annua* and *Phalaris minor* present in wheat fields and *Parthenium hysterophorus* and *Lantana camera* in non-cropped areas spread in India rapidly due to non-implementation of weed laws by government whole heartedly.

Weed laws helps in protecting the cultivators from using contaminated seed or mislabelled and also prohibiting legally the entry of noxious weeds.

10.7 Cultural management

Cultural practices like land preparation, optimum seed rate, crop density, timely weeding and fertilizer management, boost

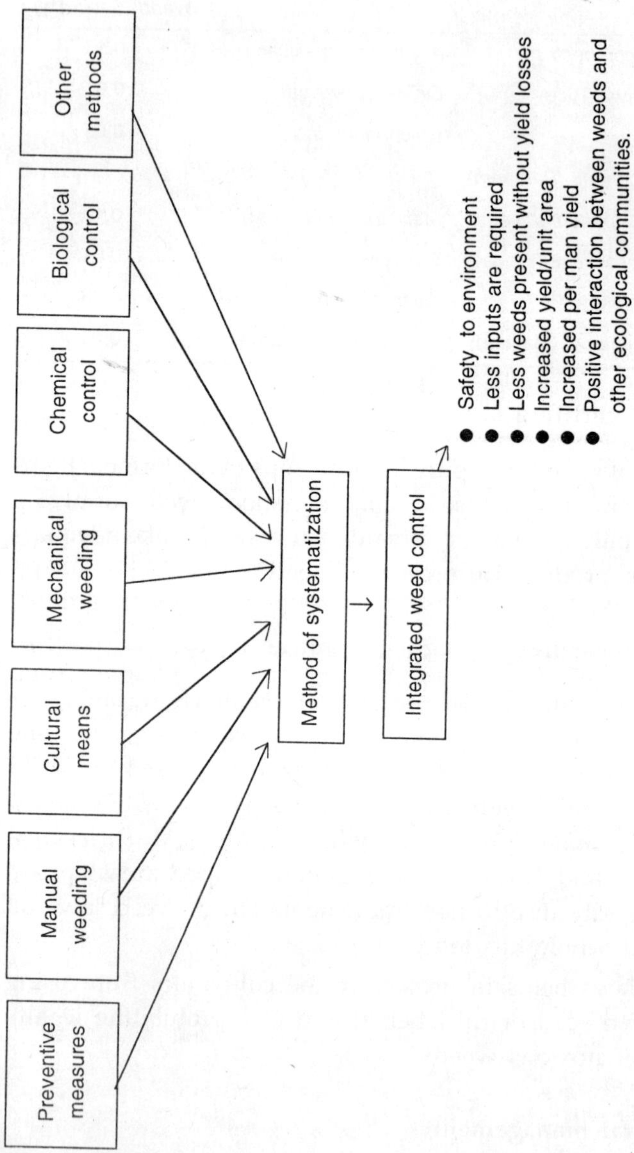

Fig. 10.1 : **Conception of Integrated weed management**

the crop growth and enhance the competitive ability of crops with weeds. These measures can substitutes chemical weeding to a great extent.

10.8 Suppressing weeds through cover crops

In north-east India, cover crops are grown to suppress weed growth in some of the plantation crops. These living mulches effectively control the weeds as inter crops. The low growing crops such as cow pea (*Vigna sinensis*), mungbean (*Phaseolus radiata*), sweet potato (*Ipomea batatus*), Egusi melon (*Citrullus lanatus*) appear to be suitable as smothering crops without competeting with main crop. Dense populations of fast growing legume cover crops such as *Centrosema pubescence, Arachis repens* and *Psophocarpus palustris* are sown as initial cover crop after establishment of main crop. Glyphosate or Paraquat is sprayed to clear 20 cm strips in which crops such as maize, sorghum and pearl millet can be planted. If the legume intercrops grown vigorously to complete with main crops, spraying with growth regulators such as GA and IAA arrest them for topping the crop. In widely spaced crops such as cotton and sugarcane and crops like coriander, *amaranthus gangetica, palak* and other leafy vegetables can be raised to arrest the late emerging weeds.

10.9 Using weed wastes as mulch

A plethora of weed species occupy wastelands and waterbodies. Some of the major weeds occupying the non-cropped areas in India are *Eupatorium odoratum, Mikania micrantha, Lantana camara, Ipomoca carnea,* and *Parthenium hysterophorus* which produce enormous quantity of biomass. This plant waste is not useful for firewood and other purposes but can be used as a mulch material in wide spaced crops and as a source of nutrients. These mulches not only supply nutrients to the crop, improve the soil fertility and moisture holding capacity of soil. Some of the weeds like *calotropis, lantana* and *Sesbania speciosa* can be chopped and incorporated in puddled soils before planting rice.

This weed waste mixed in the soil reduces the growth associated weeds significantly.

10.10 Luxurious growth of aquatic weeds

In waterbodies like drainage canals, ditches and irrigation canals, water hyacinth and salvinia are grown luxuriously. It has been reported that water hyacinth occupy 292 thousand hectares in India and has a huge potential of mulch material in plantation crops like coconut, oil palm and cocoa etc. The biomass production of water hyacinth is nearly 250 t/ha and suppress the weed germination when used as a mulch in inter-row area of orchards.

10.11 Growing of economical weeds as intercrops

Several weeds have an economical value. In rural areas, some of these weeds are used as pot herbs, forage, medicinal and dyeing purposes. These weeds have a beneficial effect on crop rather than harm. Some of the weeds like *Amaranthus, Sour sorrel, Chenopodium album, Oxalis spp., Phyllanthus articularis, Leucas linifolia, Ageratum* spp. are used as green vegetables, medicines and used for other purposes. Unless these weeds are high densities, the yield losses of crops are meagre. The fast growing useful weeds with high suppressing ability can decrease the competition of harmful weeds to a great extent. These techniques can be followed in intercropping systems. Some studies are initiated at ICRISAT, Hyderabad to determine the production economics of farming systems where weeds are included as intercrops as opposed to weed-free crop production systems.

10.12 Competitive cultivars and agronomic techniques

Wild plants and traditional cultivars of crops are good competitors with weeds. However, man has long been developing varieties based upon his economic importance. Recently crop breeding activity is merely concentrated on high yielding potential and multiple pest resistant varieties. As a result of these, present varieties are not tolerant to weed competition and yield losses are great when there is severe weed infestation. For example, in high yielding short standard rice varieties the problem of barnyard grass (*Echinochloa colona* and *E. glabrescens*) is increasing day-by-day. In dwarf HYV wheats, weeds like little canary grass (*Phalaris minor*), wild oats (*Avena fatua*), *Avena ludoviciana*, and *Convolvulus arvensis* becoming problem, which were earlier smoothered by tall traditional cultivars. The weeds *Celosia argentia*, and *Cenchrus biflorus* are posing threat in newly introduced maize varieties. Similarly in pearl-millet and finger-millet fields *Euphorbia hirta* and *Heliotropium subulatum* infestation are becoming dominant weed communities after the cultivation of HYV.

10.13 Resistant crop cultivars

In sorghum, a root parasite *striga* may pose serious problem till now. However, extensive research was done for screening striga resistant varieties in Africa and India. Sorghum varieties SAR-1, SAR-2, SAR-16, ICSV 1007BF, N-13, N-555 and *bajra* variety TF-23A are found to be resistant to striga in Indian subcontinent. Breeding of resistant variety is the best approach to control plant parasites, however, the success is scarce in identifying *Orobanche* resistant sunflower varieties even after strenous efforts.

10.14 Time of seedbed preparation

If the land is kept fallowing without any crop cover, the weed seeds germinate, grow and complete their life cycle successfully. The matured seeds are shed and perennial organs establish in the fields and become a problem for future sown crops. Time of land preparation is most important for proper crop establishment in rainfed agriculture. Delay in land preparation during monsoon season encourage the weed growth. Deep ploughing with inverted plough during hot summer expose the weeds to burning sun which results in killing of perennial weeds. Significant reduction was found in weed densities of

Cyperus rotundus and *Cynodon dactylon* when land is prepared during summer.

10.15 Stale seedbed technique

A stale seedbed technique involved destroying of successive flushes of weeds before planting any crop. This method is very good where weed seed bank is high. Sparce weed population may emerge in the succeeding crop due to depletion of weed seed bank. The weed seed which are present in the favourable zone are killed effectively. Where light showers occur for an extended period before onset of actual monsoon. In other seasons light irrigation should be given periodically alternated with ploughing for killing germinated weeds. This technique has been used successfully in countries like Philippines, Sri Lanka and India. Some of the advantages due to stale seedbed technique are

- Easy to control broadleaf weeds.
- Seed bank of weeds are drastically reduced.
- The crop establishment is very good in dry sown crops.
- The cost of weed control is nominal.
- Perennial weeds will be eliminated due to continuous disturbance of soil.

This seed bed technique has no advantage if the crop sowings are delayed during monsoon season.

10.16 Adjustment of planting dates

In every agro-ecological zone, there is a peak period of germination of weed seeds. This period is usually concides with the onset of monsoon and optimum time of crop planting. If the weed infestations are heavy, crops can be made to escape early weed competition by planting them earlier or later than their normal planting. This planting is also depends upon soil moisture available. Rainy season crops like maize, sorghum and cotton may avoid weed competition if they planted 15 days before the break of monsoon by providing light irrigation. Crop seeds germinate in weed-free environment and by the time weeds emerge with onset of monsoon, the crop seedlings grown up very well. The most problematic weed, small canary grass in wheat fields can be avoided if sowings are delayed but at the same time yields are also in decline. Slight delay in sowings of chickpea, lentil, fababean is useful in decreasing the incidence of *Orobanche* parasite.

10.17 Crop density and geometry

Crop density and planting geometry influence the canopy structure and determine the smothering capacity of crop on weeds. Heavy seed rate and narrow spacings enable the crop to cover the vacant space and reduce the weed density and biomass. Significant reduction in weed growth was noticed with the increasing maize population from 60,000 to 90,000 plant/ha through adopting narrow spacing. Similarly weed biomass production can be reduced by increasing plant population in soyabean fields. Narrow spacing cut off the solar radiation before it reaches the soil and discourage weed growth.

In wheat crop, close row spacing was highly effective in suppressing the weed population of little canary grass. Bidirectional sowing of wheat (with half the seed rate in each direction with rows crossing at right angles) reduces wild oat weed growth and gives good grain yields over unidirectional sowing. Skip row also (1 row skipped for every 90 cm or 100 cm sowing) also reduce weed growth compared to parallel rows.

10.18 Intercropping

The weed growth is heavy in sole cropping than intercropping systems. Intercrops usually suppress weeds by occupying the niches of weeds. Intercrops gives an opportunity to utilise crops themselves as tools of weed management. The troublesome perennial grasses and sedges are eliminated over time in these systems with little inputs. Maize intercropped with mungbean suppress the weeds like *Cyperus rotundus* and *Cynodon dactylon* successfully with enhancing competitive ability. Similarly sorghum based intercropping system keep down the dry matter production of weeds by 25 per cent than single cropping system. Intercropping of sorghum with groundnut and mungbean, growing of cowpea and blackgram in pigeon pea interrow spacing not only reduce weed growth but also enhance crop yields due to less incidence of pests. Intercrop of soyabean in maize fields reduce the weed population drastically due to coverage of inter-row area with crop canopy.

10.19 Proper water management

Good water management in crops reduce the weed infestation in any season. High soil moisture at initial crop growth stage encourage the weed emergence. Numerous weeds emerge when frequent irrigations are given to the crop. Drying the upper soil layer reduce the weed growth. Submergence of soil at a depth of 5-15 cm depth drastically reduce the grassy and sedge weeds in rice fields. There is a negative correlation between flood depth and weed population in rice fields. Alternate wet and dry water regimes results in emergence of weeds conspicuously. Under the unweeded conditions, yield losses are more under adequate irrigation conditions than under moisture stress conditions.

10.20 Minimum tillage

Ploughing is essential for better physical condition, for the germination and growth of crop. However the degree of land preparation or tilling is mainly dependent on nature of soil, crop and rainfall pattern. Complete tillage is disastrous in case of no rain follows after ploughing because it may cause moisture losses through evaporation. In no tillage or zero tillage plots, grassy weeds predominate on a long run. Sedges, particularly *Cyperus rotundus* proliferate in ploughed plots. Minimum tillage usually make a way to regenerate perennial weeds like thatch grass and *Agropyron repens* etc. The non-success of zero tillage is attributed to build up of perennial weeds and hardening of soil.

10.21 Rotation of crops—the best way to avoid troublesome weeds

Effective weed control is the first benefit of crop rotation. Most of the troublesome weeds dominate in fields when the same crop is grown continuously over the years. When crop rotation is followed with different cultural practices their life cycle and reproductive cycles of weeds break down and their emergence is prevented. Some weeds are associated with certain crop sequences. Certain parasitic weeds are host specific and growing non-host crops alternatively, these weeds automatically vanished. Some parasitic weeds can be reduced by sowing trap crops or by catch cropping. Trap crops stimulates germination of parasite weed seeds but these crops are not parasitised.

10.22 Effective cropping system

Cuscuta parasites are mainly grown on legume crop hosts. Growing of cereals in these fields for a while reduce the parasitic incidence. Raising trap crops like sunnhemp, sorghum, chillies, linseed, sesamum, cowpea and paddy drastically reduce the orobanche incidence. Flooded paddy effectively kills the parasite weeds due to creation of anaerobic environment. Orobanche incidence is minimum when the tobacco and vicia beans preceded by groundnut, flax, sorghum and fenugreek. The crops such as sunflower, cotton, groundnut, castor, linseed, cowpea and beans are good trap crops for striga parasite. Rotation of sorghum with these crops controls the weed effectively. Red rice weed problem may be reduced when the rice is rotated with soyabean for every two years. The most obnoxious weed, *Phalaris minor* is very high in rice-wheat, and sugarcane wheat rotations but less in maize-wheat rotation.

10.23 Method and time of fertilizer application

Fertilizer not only enhance crop yield but also encourage weed growth if it is not applied at proper time. Irregular applications without understanding the crop requirements, promote weed growth conspicuously. Several weed species are strong competitors and absorb more N, P, K and other micronutrient than crop plants at all stages of growth. Correct time and method of fertilizer application prevent weed proliferation and maximise the crop growth. Fertilizer should not be applied prior to weeding. Nitrogen application should be delayed until the nitrogen absorption of weeds has slowed down.

Broadcasting of fertilizer on entire field stimulates weed growth whereas band placement or sub-surface applications decrease the weed infestation. High rates of fertilizer applications also promote weed growth rather than crop development. Optimum fertilizer rates at appropriate time enhance the fertilizer use efficiency by discouraging weed growth.

10.24 Weed control through bio-agents

Controlling of weeds by utilization of bioagents like plants, insects and microorganisms without injuring desirable crop plants is called biological weed control. The weed populations can be suppressed to a below threashold level through bioagents without reducing the crop yields. Several bioagents like nematodes, pathogens, parasitic plants and insect pests do the weed killing on certain weed species only. Large scale weed infestation of single weed species can be suppressed by selected bioagent. This situation occurs in forests, range and grass lands and large water ways. Biological control is easy when the weed has no natural enemies in the newly introduced areas.

Some Mycoherbicides used for the control of weeds

Weed species	*Microbial species*
• *Aeschynomene virginica* (Northern joint vetch)	*Colletotrichum gloesporioides*
• *Cuscuta* spp. (Dodder)	*Colletotrichum, gloesporiodes* sp. *Cuscutae*
• *Clidemia hirta* (Kosters curse)	*Colletotrichum gloesporides* sp. *Clidemiae*
• *Orobanche* spp. (Broom rape)	*Fusarium Oxysporum* Var. *Orthoceras*
• *Diosphyras virginica* (Persiaonion)	*Cephalosporium diosphyri*
• *Morreria odorate* (Strangle vine)	*Phytophtora citrophtora*

The bioagents or insect's should have host specificity on weeds. They should have ability to kill weeds or destroy its reproductive organs. The bioagent should have hardiness to tolerate environmental conditions in which weed is growing.

11

ALLELOPATHY-A BIOLOGICAL WARFARE

'Desperate diseases require desperate remedies'

—English proverb

The concept of allelopathy is ancient in agricultural science. Theophrastus (300 BC) referred *"phytotoxicity"* among plants in *Enquiry into plants*. Democritus noticed the use of naturally occurring plant products as a practical method of controlling weeds. He also reported that trees could be killed by treating their roots with a mixture of lupine flowers soacked in hemlock juice. Plinius (AD 1) described the apparent allelopathic interactions in his book *Natural History*. He reported the use of fenugreek, chickpea and barley to scorch up or destroy maize fields.

11.1 Allelochemicals

The word *allelopathy* was coined by H. Molisch in 1937 from *allelon* ("of each other") and *pathos* ("to suffer") = although the definition covered both detrimental and beneficial biochemical interactions among all classes of plants, including microbes. Later Rice (1984) defined allelopathy as *"any direct or indirect harmful or beneficial effect by one plant on another through the production of chemical compounds that escape into the environment"*.

Allelopathy refers to the study of the production of allelochemicals, mostly secondary metabolites, by a plant that an either harm or benefit another plant. Chemicals with allelopathic potential exist in virtually all plant tissues, including leaves, flowers, fruits, stems, roots, rhizomes and seeds. These chemicals are released by the processes of root exudation, leaching, volatilization and decomposition of plant tissues. Biochemical interactions takes place when allelochecals produced by one plant escape into the environment and influence the growth and development of another plant.

Although the postulated mechanisms by which weeds affect crop growth is not known fully. But it occurs in natural plant communities.

11.2 Definition

Akobundu (1987) defined that allelopathy is the detrimental effects of chemicals or exudates produced by one living plant species on the germination, growth and development of another plant species or microbes sharing the same habitat.

A set of plant interactions should meet the conditions for allelopathy

- The inhibitional effect of one plant (donor) resulting from the interaction with another plant (receptor) should be traceable to chemicals introduced into the environment by a living donor plant.
- The detrimental effect can be shown when no growth factor that plants compete for, is in a limiting quantity.

There is another school of thought about allelopathy. Putnam (1988) is of opinion that allelopathy indicates positive plant response mediated through chemicals produced by another plant, not only negative ones.

11.3 Allelopathic interactions

The allelopathic interactions derive from the productions of secondary metabolites by higher plants which induce a wide array of biological effects.

The term "allelopathy" has undergone several changes over time (Molisch, 1937 to Rice, 1984). The allelochemicals causes both inter and intra-specific interactions between organisms were mediated through a chemical process. The definition accepted by International Allelopathy Society (IAS) considers *all biochemical interactions between living systems – where plants, algae, bacteria and fungi are included - and their environment. The allelopathy defined by IAS in "any process involving secondary metabolites produced by plants, algae, bacteria and fungi that influences the growth and development of agricultural and biological systems".*

Both *"allelochemicals"* and *"bio-communicators"* covers a range of chemicals used by organisms to exchange information, from pheromones and allomones through poisons, toxins, deterrents, phytoalexins, or allelopathic agents. There chemicals govern most organism behaviour from defence attack to social conduct with warning signals.

11.4 Allelochemicals versus Competition

Many weed scientists have misused the term competition to describe interference of weeds in crop growth. In fact interference involves both resource competition and allelopathy. But several times, they ignore allelopathy and attribute reduction in crop growth is only due to resource competition. In fact, in mixed stands of weeds and crops, several mechanisms of interference are probably working simultaneously or sequentially.

In many field trials, the decline in crop yields was assumed due to competition and allelopathic effects are ignored. However, interference also occurs when the growth factors are not limited.

Nevertheless, attempt should be made to confirm allelopathy by alleviating competition by :

- Identify the effect of aqueous extracts of donor plants on the receptors under controlled conditions in petridishes.
- Growing of donor and receptor plants in pots in the green house under simulated field conditions providing all the essential growth factors.

- Culturing receiver plants in pots containing residues of donor plant under natural conditions.
- The alleviation of suppressing/promoting effect with activated charcoal will give fair assessment of allelopathic involvement due to the adsorption of allelochemicals from the medium by charcoal.

11.5 Ways of entrance of allelochemicals

Allelochemicals are introduced into the environment in several ways. Allelochemicals are usually exuded by living roots, leached from leaves, stems and roots or released as volatile gases into the atmosphere. In addition to chemicals leached from leaf litter on the soil surface, they are also released from dead root tissues present in the subsurface soil. The soil microbes play toxic compounds while decaying roots. The possible mechanism by which a toxic chemical from a donor plant prevents normal growth in a recipient plant is by blocking of growth factors.

11.6 Allelopathic potential of weed species

Allelopathy is existent in the plant ecosystem and it occurs widely in natural plant communities. It is a mechanism by which weeds affects the crop growth. It is probably an important factor in maintaining the present balance among the various plant species. The vegetation under black walnut (*Juglans nigra*) was usually very meagre than other shady trees because the walnut leaves contains a chemical inhibiting agent *Juglone*. The chemical was washed off from the leaves to the soil through rain water. The juglone is also proved to be powerful toxin on tomato plants. Root exudates of thistles (*cirsicum* sp.) injure oat plants, while exudates of *Euphorbia* sp. and *Scabosa* sp. inhibits the growth of flax crops. Aqueous extracts of field bind weed (*Convolvulus arvensis*) and canada thistle (*Cirsium arvense*) inhibits germination and growth of wheat and barley crops.

Several weed species such as *Portulaca oleracea, Poloygonum pennsylvanucum, Datura stramonium, Digitaria sanguinalis*

Chenopodium album, Abutilon theophrasti, Echinochloa crus-galli and *Amaranthus retroflexus* are found to be produced inhibitory substances in different crops such as cabbage, radish and pepper.

11.7 Depressing effect of exudates

Quack grass (*Agropyron repens*), a problematic weed mainly in temperate climate is an excellent example of allelopathic effect of a weed on crop plants. Competitive action of quack grass was partly the result of toxious produced by its roots. This pernnial grass seriously depresses yield in cereal crops and extracts from the plant are found to be toxic to a number of crop species. The extracts of 15-30 old seedlings show the same toxicity similar to leaf extracts in maize and other crops. The chemical compounds exuded by quackgrass rhizomes interfere with the uptake nutrients by maize plants. Abundance of moisture and nutrients do not alleviate the systemeic depressing effect.

Similarly thatch grass (*Imperata cylindrica*) exhibit allelopathic effect on both crops and weeds. It often develops monospecies stands. Allelopathic effects might play a great role in suppressing the other weeds in non-cropped areas. Thatch grass successfully inhibits the emergence and growth of broad leaf weed *Borreria hispida* by exuding allelochemicals from its rhizomes. Gaint foxtail (*Setaria faberii*) infestations depress the maize yields because of exudates from roots and leachates of dead roots.

Perennial weeds are more likely to develop allelopathic reactions of importance in interplant relationships. As the life-span of annual weeds are so short, they are not able to produce and release enough toxic chemicals during their life time.

11.8 Crop residues and mulches

Residues of wheat, rye, oats, sorghum and barley are very effective in reducing weed populations in vegetable crops through allelopathy. Sorghum residues are highly effective in suppressing weeds to an extent of 90 per cent. Wheat crop residues produce phytotoxic ferulic acid and suppress prickly sida (*Sida spinosa*) and tall morning glory (*Ipomoea purpurea*). Bacteria present on the surface of prickly sida seed metabolise ferulic acid to 2-methoxy-4-ethyenylphonol which is highly toxic. Living rye (Secale cereale) and its residues provide nearly 90 per cent of weed suppression in several agroecosystems. Rye mulch can be used successfully in minimum tillage or in no till cropping system for controlling weeds. Rye grass inhibits the biomass of weeds such as common ragweed (*Ambrosia artemissifia*) and red root pigweed (*Amaranthus retroflexus*). This was ascribed due to presence of allelochemics like B-phenyllactic acid (PLA) B-hydroxybutyric acid (HBA) and 2, 4-diphydroxy-1-4-benzoxazin-3-one (DiBOA) Diboa is more phytotoxic to monocot weeds. The obnoxious butter cup weed (*Ranunculus arvensis*) has a severe inhibitory effect on wheat growth in mid-hill conditions of Himachal Pradesh. If the linseed crop is grown along with wheat the inhibitory effect will be alleviated.

11.9 Allelopathic suppression of weeds by crops

The allelopathic suppression of weeds by crops is not well documented in literature. The research work on this aspect is also very limited. When oat germplasm (3000 accessions) were screened for the production of allelochemic *scopoletin,* some accessions exuded upto three times as much as a commercial cultivated variety. The scopoletion is most active in killing the wild mustard (*Brassica kaber*) Some cucumber varieties were very effective in suppressing wild prosomillet and barnyard grass (*Echinochloa crus-galli*). The seeds of *Beta vulgaris* release allelochemics that suppress *Agrostemma* spp. weed. Some crops like maize (*Zea mays*) and lupine (*Lupinus albus*) inhibit weeds to some extent through allelopathy. The crops such as white mustard (*Brassica hirta*) and prosomillet (*Panicum miliaceam*) supperess the weeds more than 75 per cent through allelochemicals. Native sunflower (*Helianthus annus*) also have a ailelopathic weed suppression but in commercial varieties it is absent sowing of *Cassia scricea* along with forage *Stylosanthes hymata* in the field drastically reduce the population of congressgrass (*Parthenium hysterophorus*). Sweet potato also showed inhibitory effect on yellow nutsidge (*Cyperus esculentus*).

11.10 Exploitation of allelopathy through breeding

Superior weed-suppressing varieties have been reported in cucumber (*Cucumnn sativus*), oats, sunflower (*Helianthus annus*), soybean (*Glycine max*), and sorghum. One approach of weed management strategy to screen for allelopathic types in germplasm collections of crops. This character could be incorporated into a cultivar by conventional plant breeding or other genetic recombination strategies. But usually plant breeders are reluctant to cover more than a few manageable objectives when developing a crop cultivar and allelopathic potential is not enough to be high priority. There is a trade-off between the ability of a plant to fight pests including obnoxious weeds and its ability to expand plant parts of economic importance. The significance of available wild types of cultivated species with allelopathic traits that can be transferred to desirable varieties. Unfortunately, very little effort has been made to develop crops with allelopathic potential by crossing current crop cultivars with wild relatives.

11.11 Herbicides from plant origin

Recently interest is stimulated on the use of microbes as a potential source of natural herbicide products. Taxonomically, actinomycetes, and unicellular microorganisms produce wide range of antibiotics. Interestingly these organisms lies in between true fungi and true bacteria. Actinomycetes are found abundant in compost pits and in well aerated soils. These tiny organism produce herbicide compounds during the metabolic activities. Some of the products are discovered recently.

The outstanding example of such herbicide possessing natural product is cinmethylin. A portion of the molecule is *cineole,* a terpene contained in high concentrations in some desert vegetation. If the extraction of cineole from natural sources or through synthesis is economically viable, the industry of herbicide products will be greatly revolutionised. Plants can be genetically manipulated to enhance the yield of useful secondary metabolites. These secondary metabolites can be used in the same way as synthetic herbicides. In that case, several wild plant species may get economic importance for metabolite production and extraction.

11.12 Positive effect of non-crop species

Few non-crop species can release substances which promote growth of certain crop plants. Beneficial effect of cultivated flora on crop yield is claimed in various agroecosystems existed in the literature. In pearl millet (*Pennisetum typhoides*) fields, the presence of weeds particularly *Celosia argentea, Borreria articularis, Arnebia hispidissima* enhance the yield significantly. The phenomenon is not get known fully. Stimulation of growth may not be caused by positive allelopathy. Other mechanisms like improved fertility, pest control and availability of nutrients in presence of diversified weed flora results in improved crop yields.

11.13 Weed flora manipulation

Weed management can be done by harnessing less harmful weeds against more harmful weeds. Deliberate use of favourable allelopathic interaction is another strategy of improving in crop yields by decreasing the weed competition. Mexican farmers encourage some weeds species which are considered to control noxious weeds and to stimulate crop growth as well. The debris of *Tagetes patula* weed inhibits the germination of major predominant weeds of maize with little influence on yield. It is possible to establish the manipulation of weed species by means of specific allelopathic reactions of certain weed species. In future, it is possible to manipulate weed stands more efficiently in crop production through the study of allelochemics produced by various weeds.

11.14 Future needs of allelopathic potential

All crop plants produce allelochemicals, the characteristic is declining from traditional varieties to high yielding commercial varieties. Allelopathy trait was more common in the wild predecessors of cultivated crop plants which enabled them to compete with other plants to form a pure stand. It is desirable to incorporate allelopathic traits from wild relatives into more promising crop cultivars through breeding. The atrazine resistant crop conola (*Brassica napus*) developed from novel breeding in Canada. Similarly glyphosate and chlorsufaron resistant variety of tobacco and *petunia* plants were developed through genetic alteration of allelochemic synthesis in crop plants.

11.15 Autointoxication

For any crop, some quantity of unharvestable portion or stubbles is left in the soil. The crop debris is always thought to be beneficial to succeeding crops. But recent research indicates the evidence of inhibition of crop growth by the residues of preceding crop. Autointoxication is one aspect of allelopathy in which phytotoxins produced by a plant can limit the growth or population size of not only other plant species but also of its own species as well. Residues of timothy, tobacco, rye and barley that remained in the soil after harvest, released great amounts of phytoxic substances after 1-2 months of decomposition. These toxins reduce the germination, respiration and growth of different kinds of plants.

11.16 Sick soils

In some soils, stunted growth of crop plants was common even at high cultural regime. They are called sick soils. These sick soils contains phytotoxins that produce a detrimental effect on some other plants or suppress the growth of same plants from which phytotoxins are derived. Indian farmers experience growth inhibition of sugarcane ratoon when the previous crop debris is decomposed in soil. Similarly farmers in China and

Taiwan have always left huge amounts of paddy straw after harvesting. The rice residues submerged in soil could release phytotoxic substances during the decomposition. Organic compounds released during decomposition release phytotoxic substances can be detrimental to the rice seedling growth. This results in decline in productivity of second season rich crop compared to first crop. This inter-specific interaction is termed as autointoxication. The phytotoxins produced during decompositions of rice residues are vanillic, p-hydroxyenxoic, p-Coumaric, syringic, O-hydroxyphenylacetic, and ferulic acids. The autointoxication mechanism of rice is thought to be influenced by environmental factors, such as temperature, light, soil texture, nutrient availability and degree of water drainage. Higher concentration of chlorogenic acid was found in the extracts of sunflower plants grown under phosphate deficiency than with P-sufficiency plants.

11.17 Production of toxins

Usually aquatic plants may produce shikimic acid as an adaption to avoid accumulating toxic intermediates such as ethanol. A substantial quantity of shikimic acid could accumulate in paddy soil because rice plant is considered as semaquartic plant. Sikimic acid is a key precursor for synthesizing phytotoxic phenolics such as P-hydroxybenxoic, P-Coumaric, vanillic and ferulic acids and also Indole-3-acetic acid which is a growth stimulator.

The root development retards by phytotoxins present in the soil-residue mixture and this suppression can be retarded by applying ammonium-N.

11.18 Allelopathic effect of green manuring

Most of the developing world apply green leaf manures to cropped fields to compliment chemical fertilizers. This is the way of converting some undesirable plants rich in nitrogen into organic energy by incorporating them into the soil. Incorporating some organic wastes as green manure during the process of decomposition release certain substances that can inhibit germination and development of various weed species.

Wild sage (*Lantrna camara* L. Var. *Aculeata*) a beautifully coloured flowering plant introduced in India around 1810 for ornamental purpose in private gardens and front yards. It has spread alarmingly in Himachal Pradesh, Andhra Pradesh, Madhya Pradesh, Uttar Pradesh, Punjab and Haryana etc. This obnoxious weed poses a serious problem for flora and fauna because of its toxic substance *Lantadene A*. It also contains some allelopathic compounds which inhabit plant growth. *Lantana* extracts usually suppresses germination of associated weeds like *Panicum, psilopodium, Echinochloa colonum, Digitaria sanguinalis, Morrenia odorata* (Milkweed vine), *Abutilon theophrastix* (Velvet leaf), *Cyclossus dentatuts* (Cyclofern), *Commelina benghalensis*. The green twigs of *Lantana* on dry weight basis are reported to contain more than 2.0 per cent N and 1.5 per cent P but poor in K. It would appear that *Lantana* has some useful allelopathic compounds (hormonal substances) that increase tillering, chlorophyll content and photosynthetic activity in rice crop. When incorporated at 10t/ha in rice fields, it adds 80 kg/N/ha and improve the yields significantly.

12

BARNYARD GRASS—MIMETIC IN RICE FIELDS

Although some geeses have white feathers, they are not swans —*Spanish Maxim*

Introduction : The barnyard grass in a notorious mimetic weed, found both in terristrial and aquatic habitats. About 40 species of *Echinochloa* grasses are found in various field crops and non-cultivated areas. These weedy grasses are well distributed in Asia, Africa, North America, South America and Australian continents on the globe. Majority of the species are mesophytic but equally grows well in marshy habitats because of their hydrophilus nature. They compete with crop plants for nutrients, space, solar radiation and water and adversely affects the micro-climate around the plant.

At the juvenile stage, it is very difficult to distinguish between barnyard grass and cereal seedlings because of their close resemblance of morphological traits. Such type of morphological mimicry helps them to avoid removal from manual weeding. That is why, barnyard grasses are called *mimicry* weeds or mimetic weeds, especially in cereals.

12.1 Notable characteristics of *Echinochloa species*

12.2 Annuals

12.2.1 *Echinochloa oryzoides*

- Spikelets broadly ovate
- Inflorescence hanging almost horizontally at the time of maturity.
- Awned spikelets
- The length of awns is about 50 mm.
- Caryopses ovate
- Embryo length is about 0.7 mm

12.2.2 *E. crus-pavonis*

- Inflorescence strongly drooping at maturity
- Spikelets crowded with short and curved awns.
- The length of awns ranging from 3-10 mm

12.2.3 *E. frumentaceae*

- Spikelets pale green even at maturity
- Spikelets awnless
- Whitish caryopses

12.2.4 *E. utilis*

- Spikelets brownish at maturity
- Spikelets awnless, occasionally awned
- Brownish caryopses.

12.2.5 *E. zelayensis*

- Main axis and branches of inflorescence are not having long bristles.

- Spikelets finely pubescent
- Palea of lower floret absent

12.2.6 *E. phyllopogon*

- Spikelets ovate - elliptical
- Inflorescence is erect at maturity
- Spikelets awned, sometimes awnless.
- Lower lemma convex shaped, shiny and rigid
- Tufts on hairs present at collar region of leaves
- Oblonged caryopsis

12.2.7 *E. subverticillata*

- Spikelets had short bristles
- Whorl shaped bristles present at the base of inflorescence
- Lower floret staminate

12.2.8 *E. callopus*

- Spikelts with small globular callus below glumes

12.2.9 *E. ugandensis*

- Narrow leaf blades which are not wider than 5 mm
- Panicles linear and erect

12.2.10 *E. elliptica*

- Leaf blades are broad and wider than 10 mm
- Bristles are long and numerous arising at nodes of inflorescence
- Spindle shaped panicle (15 cm long)
- Spikelets narrow and elliptical
- Awns of lower lemma about 30 cm long and second glume up to 10 cm long

12.2.11 *E. turneriana*

- Bristles are absent along main axis
- Panicle is linear and narrow
- Spikelets broadly ovate and sometimes ovate-ellipticle.

12.2.12 *E. glabrescens*

- Spreading habit of tillering
- Found on banks and fallow-lands
- Spikelets 3-3.5 mm long
- Lower lemma convex shaped, hard and shiny

12.2.13 *E. crus-galli* Spp. *crus-galli* var. *crus-galli*

- Spikelets broadly ovate, short-awned or long awned.
- Caryopsis ovate shaped.
- Panicles are erect, variable in length.
- Pyramid shaped panicles, sometimes nodding.
- Long panicles, after having secondary branches on lower primary ones.

12.2.14 *E. crus-galli* sp *crus-galli* var. *hispidula*

- Spikelets ovate-elliptical.
- Long awns present sometimes short.
- Panicle branches often whorled, erect, except lower most ones.

12.2.15 *E. crus-galli* sp *hispidula* var. *Austro-Japonenes*

- Spikelets ovate-elliptical with short awns.
- Inflorescence short, Close and less erect branches.
- Inflorescence branches sometimes very long upto 28 cm.

12.2.16 *E. crus-galli* sp *crus-galli* var. *praticola*

- Spikelets irregularly arranged.
- First glume about one-third length of spikelet.
- Brownish caryopses.
- Long bristles along main axis.
- Branches are present on inflorescence and sometimes absent.

12.2.17 *E. colona*

- Spikelets arranged in rows.
- First glume is half with the length of spikelets.
- Whitish caryopses
- Long bristles are occasionally present along the branches.

12.2.18 *E. walteri*

- Length of spikeltes is about 5 mm.
- Bristles are prominent on spikelets
- Prominent bristles are present on leaf sheath.
- Leaf sheath has light pubescence.

12.2.19 *E. rostrata*

- Spikelets having short curved awns.
- Spikelets are crowded in clusters along the main branches.
- Inflorescence arranged on stout main axis and 40 cm long.

12.2.20 *E. senegalensis*

- Spikelets crowded in regular rows.
- Panicles are 40 mm long with erect branches.
- Branches closely appressed to main axis.

12.2.21 *E. brevipedicellata*

- Spikelets are irregularly grouped along the branches.
- Panicle is 30 mm long and branches are some spreading.

12.2.22 *E. jubata*

- Spikelets are elliptical with 15 mm long awns.
- Panicles are dense with short branches.

12.2.23 *E. oplismenoides*

- Spikelets are elliptical with fine, sparse bristles.
- Panicle erect with narrow branches.
- Awns are 10 mm long.
- Palea of lower floret is sometimes absent.

12.2.24 *E. inundata*

- Spikelets are 3.5 –5 mm long.
- Panicles are linear and 3-4 mm long.
- Branches are close and 6 cum long.
- Lower floret sometimes staminate.

12.2.25 *E. telmatophila*

- Spikelets are about 4.5 mm long.
- Leaf sheath glabrous.
- Spreading bristles on spikelets.

12.2.26 *E. macrandra*

- Anthers are 2.8 cm long.
- Plea and lemma are same length.
- Lower floret staminate.

12.3 Perennials

12.3.1 *E. stagnina*

- Spikelets are 3-5 mm long, arranged in 3-4 branches.
- Lower palea present.

12.3.2 *E. picta*

- Spikelets having 0.5 mm long bristles.
- Uppermost leaf blades are about 10 mm wide.
- Ligular bristles are absent in uppermost leaves.

12.3.3 *E. haploclada*

- Spikelets are 3 mm long, having short awned.
- Ligular bristles are absent.
- Tail plant with close-tufted habit.

12.3.4 *E. pyramidalis*

- Plants 4 m tall with stout culms.
- Inflorescence is more than 40 cm long.
- Spikelets are crowded and finely pubescent.
- Secondary branches closely appressed to primary branches of influrescence.

12.3.5 *E. helodes*

- Spikelets lanceolate and 2 mm wide.
- Lower glume three-fourths length of spikelet.
- Lower palea present.

12.3.6 *E. holciformis*

- Spikelets elliptical, having 2 mm wide.

- Lower glume extending slightly beyond base of lower lemma.
- Lower palea absent.

12.3.7 *E. holubii*

- Spikelets often wooly ciliate
- Culms less than one metre tall.
- Inflorescence branches closely appressed to main axis and distant.
- Culms less than one metre tall.

12.3.8 *E. lelievrei*

- Spikelets 4-5 mm long arranged in 2 rows along with branches.
- Lower palea present.

12.3.9 *E. praestans*

- Spikelets finely pubescent.
- Bristles on ligule are present in all leaves.
- Culms are stout and 3.6 mm tall.
- Leaves are broad.
- Uppermost leaf blades are about 20 mm wide.

12.3.10 *E. madagascariensis*

- Culms are less than 1 mm tall.
- Leaves are narrow, about 6 mm wide.
- Hairs are present on whole plant.

12.3.11 *E. polystachya* var. *polystachya*

- Spikelets are broad-ovate.
- Awns rarely frequent.
- Leaf sheaths glabrous.

12.4 Ecology of jungle rice

Jungle rice is pernicious weed and inflicts enormous losses in various crops such as cereals, vegetables and plantation crops. It is a most notorious weed in rice and causes huge losses to rice growers in terms of labour and yield.

12.5 Biology of weed

Jungle rice, *Echinochloa colonum*; Syn. *Panicum colonum* inspite of its name, it is not a rice but an annual grass and the most persistent weed in the river deltas of peninsular India. Locally known as *ooda*, it competes with rice for native and applied nutrients, water, radiant energy and space and adversely affects the microenvironment around the plant. It also harbours diseases like sheath blight and insects like grass hoppers.

12.6 Ecomorphology

Jungle rice is a hydrophilus weed but grown equally in well-aerated soils. It may reach up to a height of 85 cm and produces about 33 tillers per plant. The root system is fibrous, found up to a depth of 15 cm. It has an enormous seed producing ability, one plant producing around 460-740 seeds.

The weed matures earlier (60 days) than rice and shatters its' seed. These seeds remain viable for about 3 years even under waterlogged conditions. They may germinate and give seedlings in dryland from a depth of 6 cm when the soil is hard and compact, but when the soil is loose and moistened, the seedlings may come out even from a depth of 12 cm.

12.7 Acts as mimicry artist

Early in the season it is often hard to differentiate between the young rice seedlings and those of weed plants due to close resemblance. The weed seedlings bear a striking resemblance to the rice seedlings, so much so that at the time of nursery pulling. Hence these weed seedlings also transferred to the mainfield along with rice seedlings during transplanting operations. Sometimes more than half of the hills are mixed with weed seedlings due to physical mimicry with rice seedlings. The removal by hand is very difficult because of its difficult identification with naked eye. Consequently they survive in the mainfield.

12.8 Some of the key points to identify jungle rice

Plant part	*Rice*	*Jungle rice*
Stalk/Stem	Contracted	Spreading, swelling at internodes.
Base of leaf blade	Ligule on lower leaf surface and auricles on both leaf surfaces	Hairless, no ligule
Leaf shath	Green	Red or purple tinge
Height	Tall	Short
Shoot	Thick, flattened	Thin, round

The weed cannot be eradicated due to its inherent hardiness, persistence even under unfavourable conditions and a prolific-seed production ability. However, preventive methods, cultural practices and chemical treatment in proper combination will provide effective weed control.

12.9 Distinguishing characteristics between two *Echinochloa* species found in tropics

Character	*E. colonia*	*E. crus-galli*
Panicle shape	Erect	Nodding is very common
Rachis branches	Simple, distant on a rachis	Branched, Close on a rachis
Spikelet	Arranged in 4 rows	Irregularly arranged
Awn	Awn is absent	Length varies with region
Stigma	Purple, blackish Purple	White/red
Anther	Purple	Brown/yellow
Culm	Spreading	Erect

12.10 The most dominant barnyard grass prevailed in various countries

Species	*Distribution*	
	Climate	*Countries*
Echinochloa colona (L) Link	Tropics and Subtropics	India, Bangladesh, Myanmar, Pakistan, Sri Lanka, Japan, and Korea.
E. oryzicola (Vasing)	Subtropics and temperate	North India, Japan, China, Korea, Italy, Sri Lanka, France.
E. crus-galli (L) Beauv	Warm temperate and subtropics	Philippines, India, Sri Lanka, Japan, Korea, Bangladesh.
E. crus-pavonis (HBK) Shult	Tropics	South American Countries like Brazil and Africa.
E. stagnina (Retz) Beanv	Tropics	India, Bangladesh, Sri Lanka, Philippine, Thailand, Vietnam, Myanmar, Mali.
E. pyramidalis (Lam) Hitchc chase	Tropics	African countries like Nigerea, Urugway, Kenya etc.
E. utilis ohwi and yabuno	Temperate	Japan, Korea, China.
E. frumentacea Link	Subtropics	India, Pakistan
E. haploclada (stapf)	Tropics	East Africa
E. obtusiflora stapf	Topics	West Africa

12.11 Distribution of barnyard grasses on the globe

12.11.1 *Asia continent*

Echinochloa crus-galli (L) Beauv. sp. *crus-galli* var. *Praticola* ohwi

Echinochloa crus-galli (L) Beauv. sp. *hispidula* (Retz) Honda

Echinochloa crus-galli (L) Beauv. sp. *hispīdūla* (Retz) Honda var. *Austro-japonensis,* Ohwi

E. *candata* Roshev

E. *frumentacea* Link

E. *colona* (L) Link
E. *glabrescens* Munro ex Hook
E. *oryzoides* (Ard) Fritch
E. *phyllopogon* (stapf) Koss
E. *pica* (Koen) Michael
E. *praestances* Michael
E. *utilis* ohwi et yabuno

12.11.2 *Africa*

E. *rostrata* (stapf) Michael
E. *rotundiflora* clayton
E. *senegalensis* Mez
E. *subverticillate* pilger
E. *stagnina* (Retz) Beauv
E. *ugandensis* snowden
E. *verticillata* Berhant
E. *bealnanensis* (A. camus) Michael
E. *brevipedicellata* (peter) clayton
E. *callopus* (pilger) clayton
E. *haploclada* (stapf) stapf
E. *holubii* (stapf) stapf
E. *jubata* stapf
E. *lelievrei* (A. Chev) Berhaut
E. *madayascariensis* Mez
E. *obtusiflora stapf*
E. *pyramidalis* (lam) Hitchc. et chase

12.11.3 *North America*

E. *holciformis* (H.B.K.) chase
E. *macra* (wreg) Michael
E. *microstachya* (wieg) Rydb

E. muricata (Beauv) Ferm

E. oplismenoides (Fourn) Hitche

E. paludigena Wieg

E. walteri (Prush) Heller

E. zelayeosis (H.B.K.) Shult

12.11.4 *South America*

E. crus-pavonis (H.B.K.) Shult

E. polystachya (H.B.K.) Hitchc var. *polystachya*

E. polystachya (H.B.K.) Hitchc var. *spectabilis* (Nees) Martinez crovetto

12.11.5 *Australia*

E. elliptica Michael et vickery

E. immdata Michael et vickery

E. kimberleyensis Michael et vickery

E. lacunaria (F. Muell) Michael et vickery

E. macranda Michael et vickery

E. telmatophila Michael et cickery

E. turneriana (Domin) J.M. black

E. crus-galli (L) Beauv-sp *crus-galli* var. *crus-galli*

13

NOXIOUS WILD RICES

Shun not the struggle, it is God's gift.

Wild rices are usually recognised as weeds in rice fields as their infestation cause several weedy problems. Their competition is highly intensive in dryland situation. The *Oryza* species which are considered as weeds are *Oryza rufipogon, O. punctata, O. barthii, O. longistaminata, O. nivara* and *O. officialis. Oryza* is characterized by the development of a pair of auricles at the base of the lamina and rudimentary glumes at the upper part of the pedicel. Some botanists re-classified *Oryza subulata* into *Rynchoryza subulata* by utilizing the above characters. Out of 20 species of *oryza,* only two are domesticated-*Oryza sativa,* which is cultivated in the Asian tropics and subtropics, and *O. glaberrina,* which is grown in African continent. Surprisingly one of the wild rices of North America, *Zizania aquatica* grown in aquatic bodies produce high quality grain and considered as a culinary delicacy. The grain fetches very high prices and rich people can afford for it.

13.1 Evolution of wild rices

Wild rices are derived from domesticated rice species as a result of introgression with wild relatives. In Asia, domesticated rice species *Oryza sativa* when naturally crossed with *Oryza rufipogon* and *O. nivara,* a series of wild rice hybrids came into exists. The African domestic species *Oryza glabrrima* when existed nearer *O. longistaminata* and *O. barthii*, swarms of hybrid wild rices with many intergrades are produced. The plant *Oryza sativa* type *spontanea* is a hybrid progency among the Asian species. These hybrids are the weeds called red rice, because they possess dominant red pericarp.

13.2 Characteristics of wild rices

- Several wild rice species like *O. rufipogon* and *O. l'ongistaminata* are perennials.
- All the wild rices resemble *O. sativa* in most of the aspects.
- They have ability to cross pollimate.
- The grains have strong and long dormancy.
- Ability to survive in soil for several years.
- Some red rices possess genotypically dominant red pericarp.
- Have a habit of easy shattering of grain.
- The red pericarp results in milled rice with a dirty whitish or yellowish colour.
- The red pigment is provided by anthocyanin pigment governed by two complementary genes, Rc, Rd and Pl.
- Some wild genes like *O. longistaminata* have a strong underground rhizomes.
- Red rices are hybrids between cultivated rice and wild relatives.

13.3 Distribution in various continents

The weediness of wild rices are attributed to it strong competitiveness with cultivated rice, more efficient propagation though both vegetative and reproductive ways. Red rice is the main problem in West Indies, Colombia, Venezuela, Surinam, Brazil, Guyana and southern states of U.S.A. Wild rices also posing problems in rice cultivated areas of Asia. High frequency of wild rices are found in Thailand, India, Bangladesh, Malaysia and Indonesia. In Europe red rice was reported in Italy and Bulgaria. Sporadical occurrences of red rice was noticed in rice growing areas of Africa, where *Oryza sativa* has been cultivated. These red rices belongs to African taxa.

In many parts of Africa, the strong rhizomatous *Oryza longistaminata* is a noxious weed and creating a lot many problems in rice cultivation.

13.4 Distribution of wild rices

Oryza longistaminata is a serious weed in Senegal, Nigeria, Mali, Guinea and Sierra Leone. It is also distributed in Uppervolta, Ivory Coast, Benin and Camerown. The wild rice *Oryza perennis* sp. *madagascariensis*, morphologically similar to *O. rufipogon* is a potential weed in Malagasy. A strong rhizomatons wild species *O. australiensis* is a potential noxious weed in Australia.

Oryza barthii is a nuisance weed is broadly distributed from Mauritania to Sudan and Zambia. It is highly problematic weed in Tanzania. This wild rice is close relative to the African rice cultigen *Oryza glaberrima*. The weed has red pericarp and shetters easily. These characters are controlled by dominant genes.

An annual *O. punctata* in a native of African continent whereas *O. officinalis* is of Asian origin. But both have small grain, 5.00-6.40 mm long. *Oryza punctata* is widely distributed from eastern side to West Africa. It shatters its grain profusely and contaminates the rice.

All wild rices have a strong dormancy and do not germinate until the favourable conditions arrive. There are pioneer vegetations that usually needs a considerable solar radiation to germinate, grow and reproduce. That is why, weed seeds will not germinate in the dark while they are buried in the mud deeply. Several wild rices prolong their dormancy under flooded conditions like *Echinchloa crus-galli* and *Fimbristylis miliacea* seeds. The seeds of *Oryza punctata* do not germinate even under saturated condition of the soil. However, some perennial wild rices like *O. longistminata* and *O. rufipogon* propagate through

Table 13.1 : Morpho-physiological characters of wild rices

Character	*Species*		
	O. sativa	*O. rufipogon*	*O. nivara*
Habitat	Domesticated culture in wet and dry lands	Stagnant ditches and seasonal swamps	Shallow ponds and drainage pools.
Growth Habit	Erectotype	Decumbent	Semidecumbent
Life Cycle	Annual	Perennial	Annual
Branching	Intravaginal	Extravaginal	Intravaginal
Spikelet shape	Variable (Bold to slender)	Slender	Bold
Apiculus	Greenish	Blood red	Purple or greenish
Pericarp	Variable	Purplish	Reddish-dirty white
Another length	Medium	Long	Medium
Awn	Slender/awnless to awned	Slender/flexuous and reddish	Stout, long, reddish
Grain dormancy	Variable (Low to high)	strong	strong
Grain shattering	Variable	Readily before fully ripe	Before fully ripe

Table 13.2 : Difference in morpho-physiological characters of some wild rices

Character	*Species*		
	O. glaberrima	*O. barthii*	*O. logistaminata*
Habitat	Paddy fields, seasonal swamps	Paddy fields, Shallow ditches	Paddy fields, waterlogged ditches
Life-span	Annual	Annual	Perennial
Rhizomes	Absent	Absent	Present
Panicle shape	No secondary branches	Secondary branches present and rarely tertiary branches	Both secondary and tertiary branches are present
Grain	Glabrous	Hispid	Hispid
Awn and its Colour	Awnless, short awned, hispid	upto 16 cm, pinkish or reddish, stiff hispidulous	upto 8 cm. rigid
Grain domancy	High	High	High
Growth habit	Semi decumbent	Semi decumbent	Semi decumbent

rhizomes and sprouted stem cuttings very rapidly under flooded ecosystem.

13.5 Control measures

13.5.1 *Prevention*

The maxim *"prevention is always better than cure"* for controlling wild rices. Rice seed should be perfectly clean while sowing in the nursery. The seed should procure from uncontaminated areas. Infestations are increasing in several Latin American countries through contaminated seed. Use of certified seed in the first step to control these weeds. In some countries such as Indonesia, certification is directed only for Barnyard grass and Cyperaceous weed seeds. Over looking the seed purity encourage the weed infestations rapidly. The wild rice *O. rufipogon* becoming problem in direct-seeded rice of Indonesia due to use of impure seed. Similary in Brazil, 30 seeds of wild rice are allowed in 1000 grain sample. In Indonesia, 10-15 per cent red rices are allowed and accept in seed lots of seed centers.

13.5.2 *Cultural technology*

- Direct sowing of sprouted seed, followed by flooding the field lessen the weed infestation. Fields should not be dried for long time.
- Through puddling and perfect levelling are prerequisites for better seedling growth of crop.
- Row planting or row sowing helps in the identification of wild rices present in between rows and can be weeded out.
- Developing and growing of rice cultivars having partially purple colour to distinguish the crop and weed helps in easy rogueing by manual labour.
- However, domestic rices can cross pollinate with wild rices after few years. As a result, the progeny of wild rices may get purple coloured leaves like cultivated rices.

- Wild rices usually matures very earlier and shatters its seed and improves its seed bank in the soil. To break this life cycle, farmers should grow short-season cultivars which comes to harvest earlier than wild rices.
- Crop rotation with other crops reduce the infestation of weed due to unfavourable environment. Using smothering and early covering crops like cowpea and sweet potato suppress the weeds even after germination.
- Deep ploughing in summer bury the wild rice seed in deeper layers of soil and prevent them from germinating.
- Buffalos, domestic hens and ducks are to be kept in the fallow fields as they graze the seed and seedlings of wild rices.

13.5.3 *Chemical control*

Use of molinate at 3-4 kg/ha selectively control red rice plant in rice fields. The herbicide antidote NA (1, 8-naphthalic anhydride) helps in prevention of crop injury from molinate. The NA should be applied as grain dressing at 0.5 per cent. The species *Oryza punctata* can also be controlled with molinate. In fallow fields, application of diuron (5 kg/ha) and dalapon (10-15 kg/ha) kills the rhizomes of *O. longistaminata* but the chemical residues persists for a long time which may injure succeeding crop. Glyphosate (2-3 kg/ha), being trans located herbicide kills the rhizomes of perennial wild rices. The residues took very short time (4 days) for degradation.

13.6 Preplanting herbicides

The herbicides alachlor (0.75 kg/ha) and molinate (1-2 kg/ha) can be applied preplanting for selective control of weed. Incorporating in the soil was highly selective. Thiobencarb (2-3 kg/ha), alachlor (0.75 kg/ha) were also effectively control in *O. punctata*. The selectivity of these herbicides can be increased if the cultivars are protected with NA. The herbicides 2, 4-D and MCPA cannot control red rice but are useful to control volunteer rice.

14

MENACE OF RED RICE

'Good wine needs no bush'

—French proverb

Red rice (*Oryza sativa/O. rufipogon*) is a serious weed in many rice growing countries. About seven *oryza* species are found rice weeds in more than 40 countries. These red rices are not most troublesome weeds in Asia-Pacific region but causing severe financial losses in rice producing states of USA and Latin American countries. Much confusion existing with regard to description of red rices, proper taxonomic classification and extent of damage due to its infestation. Usually some of the *Oryza sativa* varieties are considered as red rice weeds. In another countries *Oryza rufipogon* referred as troublesome weeds.

14.1 Characteristics of red rices

- Pigmented aleurone layer is present in the grain.
- Ease of shattering of mature seed.
- Unique seed dormancy.
- Capacity of reinfestation in rice fields.
- Stabilization of red rice ecotypes in the paddy lands.

The main problem with red rice is deterioration of grain quality in rice production. No red leaves and stem are present on red rice plant. The plant is normal like common rice variety but have reddish coloured grain because of *red aleurone layer* on head rice. The red aleurone layer reduces the grain quality greatly.

14.2 Negative side of red rice

Well documented estimates suggests that rice yield losses are as high as 65 per cent in U.S.A., Guyana and other countries. Nearly 90 per cent samples of harvested rice had more than 2 per cent red rice in southern states of U.S.A. Undoubtedly these rices do not get good price in the market. Seed sales also very difficult due to fear of red rice infestation in these areas. It is estimated that annual losses is more than 15 millions dollars in Arkansas and more than 20 million dollars in Lousiana states of U.S.A. The financial loss owing to reduced quality, is nearly as great as yield declines. Although red rice is a serious problem in few countries, other rice producing countries should not take it in lighter way. They should be vigilant and on guard to prevent it from becoming a serious problem in new areas. This applies to various states within a country as well. Farmers usually unsuccessful in preventing the spread of red rice into new production areas.

14.3 Difficult to control

The taste of red rice is perfectly good but identification in the field is difficult due to similar morphological nature. An effective red rice control programme requires more dedication than many farmers have. Integrated weed management programme should be implemented on a long range for effective control. One of the impediments to successful red rice control is short term lease programme. One year lease is very common between rice farmers and the land owners. This lease does not favour an approach to red rice control that requires a programme extending over one year. A more effective contact would be one which covered several years and detailed a crop rotation plan that specified alternate crops. Recommended herbicides should be used in rice crop. Both the farmers and land owners would benefit from the improved contractual arrangement.

14.4 Integrated system of red rice control

A successful red rice control programme involves all possible methods of preventing the introduction of red rice into new fields and reducing the infestation of red rice seed in the soil to prevent reinfestation. Cultural practices such as water management, crop rotation and seed bed preparation minimize the red rice problem significantly.

14.5 Proper water management reduce the red rice germination

Dry seeding usually result in more weed seed germination and higher infestation than wetseed bed preparation. Continuous flooding discourage the red rice problem to a great extent. Delayed flooding after germination will result in serious infestations of the weed in direct seeded plots. Wet seed bed preparation and planting in the flooded fields reduce its infestation. Keeping the soil under saturation decrease the red rice infestation than kept under dry and well aerated. Thorough puddling of land and continuous impounding water in the field minimise this problem.

14.6 Draining aggravate the problem

A drainage of a two-week resulted in a high number of red rice in the field. In some countries, the following cultural practices are used to control this weed :

- Zero or minimum tillage before flooding the field.
- Severe puddling under water.
- Using presprouted seeds for sowing.
- Managing shallow flooding to inhibit red rice germination after sowing.

The problem is that most of the farmers are reluctant for deep flooding because of the less establishment of seedlings. Seedlings under flooded conditions does not establish good root system. The seedlings tends to float in the water and is often blown to one side of the field. Draining the field after seeding and reflood it gradually as soon as the seedling initiated roots, usually one weak after transplanting. The land should be perfectly levelled to control the flood water.

14.7 Crop rotation

Rotation with upland crops effectively controls the red rice in rice production plots. Inclusion of soybean, sorghum and maize reduce the weed infestation. Two year rotation with soybean will decrease the severity of red rice problem. Application of herbicide (propazine) and tillage gives about 95-100 per cent red rice in sorghum rotation provides for tillage to stimulate red rice germination and herbicides kill them and prevent reinfestation.

14.8 Cultural means

Use of red rice free seed and clean farm equipment should be used in all field operations to free the soil of red rice and prevent infestation. Red rice seed is often carried from one rice field to another with puddling and harvesting equipment. Seed is also transported through irrigation water. All field equipment should be cleaned thoroughly when starting the operations in the new fields. Seed lodged in the mud on the tires or else where on the equipment should be removed with the help of water servicing. The air-crafts which are used for aerial seeding should also be cleaned. The backbone of red rice control is, preventation of seed is better than control.

14.9 High seedling density of crop

Heavy seed rates or high density planting of crop usually reduce the weed growth due to crop competition. High crop density drastically reduce the tillering of red rice. Doubling of seed rates from 100 kg/ha to 200 kg/ha reduce the red rice seed production by 50 per cent. Application of pre-harvest dessicant also helps in arresting the production of viable weed seed.

14.10 Hand weeding

Under the lower infestations of red rice, hand weeding with manual labour is justified when the labour is inexpensive. Red rice free grain fetch good price in the market. Sometimes it is difficult to identify red rice in the crop due to similar morphological trails. Use of purple leaved cultivars facilitate easy identification green leaved red rice while weeding. Selection of early maturing crop cultivars which comes to maturity before red rice.

14.11 Biological control

Encourage rodents, birds and domestic hens to feed on shattered red rice in the field after the crop harvest. Ducks and country hens feed these weed seeds in large quantities and reduce the infestation in the next crop. Ploughing should not be done immediately after harvest because of reinfestation problem. The shattered seed will go into deeper layers of soil and remain viable for many years. The seed germinate whenever environmental conditions are favourable.

15

PARASITIC WEEDS

Parasites are prominent examples of ingratitude in the world of plant kingdom.

Currently, parasitic weeds have been the subject of several research fora. For all groups of parasites, longer term control programmes need an integrated approach for effective control. More attention should be paid for bio-control as it is highly effective and more economical compared with physical and chemical control. Among the plant parasites *cuscuta* sp. *Orobanche* sp. and *Striga* sp. and others belonging to *Viscaceae* and *Loranthaceae* (mistletoes) family creates great menace and considered to be serious plant pests.

15.1 Dodders (*Cuscuta* sp.)

A great many species are registered in Asian and African countries and they cause problems in several agricultural crops. The stem parasite belonging to *cuscuta* have unspecialized seed biology and germinate spontaneously without the involvement of any host plant. However, most *Cuscuta* sp. produce good proportion of hard seeds which require scarification before they can germinate. This results in continuous pattern of germination over a period of years after seed shedding. It will reduce the risk of too many seeds germinating in the absence of suitable host plants. Once the seed germinated, the radicle is quite abnormal, short and swollen, growing only about one cm before withering. The seedling lives few days after germination. The seedling is unusual. In the mean time, the shoot grows out as a slender yellow stem, swinging in a wide circle until it strikes any plant or object, around which it immediately starts twining. If the host is suitable, haustoria will begin to develop on the surface of crop and penetrate into the vascular tissues of host.

15.2 Host specificity

If the suitable host is not available or contacted, the seedling will lose it capacity to parasitize after seven days. After it attaching the host, the *cuscuta* seedlings grow very speedily. The *haustoria* (parasitic roots) have been found to be potent metabolic sinks, exerting a powerful drain on the resources of the host plant, and often preventing the formation of fruits and seeds. Serious infestations may very often cause total crop failure.

15.3 Some prominent '*cuscuta*' species

15.3.1 *Cuscuta monogyna* Vahl

- Thrives primarily on woody hosts.
- The stems are thick and robust.
- Flowers 3-4 mm in diameter, arranged in elongated raceme.
- Conical shaped capsule.
- Stigma is single and seeds have 3 mm diameter size.

15.3.2 *C. planiflora* Ten

- Grows on a wide range of herbaceous crops and weeds.
- Thrives also on small shrubs.
- Stems are orange yellow in colour.
- The width of stem is slender to medium in width.
- Unlike *C. campestris* does not produce tendrils.
- The main stem twine round the host.
- Flowers 1.5-3.5 mm in very dense, compact clusters 4-12 mm across.

- Two long styles/stigmas are present.
- Capsule is spherical shaped.
- Seeds are less than one mm. in size.

15.3.3 *C. campestris* Yuncker

- Originate from North America.
- Grows on a wide range of herbaceous crops and weeds.
- Thrives also on citrus plants.
- Bright orange-yellow coloured stems with a medium width (1-2 mm)
- Flowers 2-3 mm in clusters.
- Corolla is lobed with incurved tips.
- Two styles, tipped with small but distinct spherical stigmas.
- Most widespread agricultural problem in several crops.

15.3.4 *C. chinensis* Lamarck

- Grows on a range of herbaceous hosts.
- Stems are slender with orange-yellow colour.
- Flowers 2-3 mm size.
- Tips of corolla lobes flattened.
- Had two styles.
- Spherical shaped capsule, circumcissile.
- Seeds 1-1.5 mm diameter.

15.3.5 *C. pedicellate* Ledeb

- Grows on a wide range of herbaceous crops and weeds including some grasses.
- Stems are slender.
- Flowers 2-3 mm had four corolla lobes.
- Styles form a slender-V.
- Spherical shaped capsule with 1-1.5 mm size.

15.3.6 *C. palaestina* Boiss

- Thrives mostly on shruby plants.
- Stems are very slender.
- Flowers 1.5-2.0 mm size in compact clusters.
- Four corolla lobes are present.
- Stigmas form a slender V.
- Capsules are spherical shaped.
- Seed size is about 1 mm.

15.3.7 *C. reflexa* Roxb

- Grows on woody hosts extensively.
- Stems are thick and robust.
- Flowers 6-8 mm size.
- Capsule is conical with 5-10 mm size.
- Two stigmas, tapering, forming a-V.
- Seeds size is about 3 mm.

15.3.8 *C. epithymum (L) L*

- Proliferates on a wide range of herbaceous crops and weeds.
- Stems are reddish and slender.
- Produce 2.5-4 mm sized flowers arranged in dense clusters of 10 mm across.
- Styles long and slender.
- Capsule is spherical shaped.
- Seeds have a size of 1 mm

15.3.9 *C. epilinum* Weike

- Survives and infests on flax and linseed.
- Stems are slender to medium.
- Had little branches.

- Flowers 3-3.5 mm in size and arranged in compact clusters.
- Styles form a-V.
- Capsules are spherical shaped.
- Seeds are 1-2 mm in size.

15.3.10 *C. pedicellate*

- Grows mostly on shrubs and weeds.
- Stems are slender and orange-yellow in colour.
- Similar to C. *palastina*.

15.4 Susceptible host plants

Cuscuta species have less host specificity than the root parasites. Grasses are rarely attacked but a very broad range of herbaceous plants can be parasitized. The age of host is also one of the important factor for parasitization. Older tomato plants become resistant to these parasites. Potato plants that contains more calcium in the plant sap are resistant to *Cuscuta reflexa*. Weeds are very often attacked by *Cuscuta* species than crop plants. Among weeds *Convolvulus arvensis* L. is one of the favoured hosts for this parasite. Among crops alfallfa is most frequently parasitized. Most vegetable crops are highly susceptible to *cuscuta* spp. The stem parasite, *Cuscuta compestris* is most problematic in lentil fields. Citrus crops highly susceptible to these parasites and cause heavy damages in Pakistan and Saudi Arabia.

15.5 Prevention is desirable than eradication

Once the *cuscuta* is infested in fields it is not possible to eradicate. It spreads like wildfire in the surrounding fields. The use of clean seed is vital in areas where infestation not occurred already. If it is already infested it should be prevented further spread. Alfalfa seed should be cleaned by special machinery.

Velvet rollers combined with iron filings and magnets should be used to separate crop seeds from the parasite.

15.6 Hand pulling do not resolve the problem

Farmers usually pull the parasite with hands but the problem is rarely resolved because any fragments of parasite stem left with haustoria roots attached to host tissues can regrow and reestablish its infestation. Harrowing with spike teeth harrow is practised at early stages in alfalfa to prevent initial establishment of *cuscuta* seedlings in the field. In some of the crops, it is better to burn the crop/weed mass to destroy *haustoria* present on the plant bases. Alfalfa can recover from drastic defoliation and damage to above grown parts. Large flame-throwers can be used to burnt the infected crop.

15.7 Crop rotation and cultural practices prevent build-up of seeds

Crop rotation with resistant crops like small grain crops and grasses may helps in controlling the weed. This will reduce the build-up of seed in the soil over the years. However, *cuscuta* has wide range of hosts and even some grasses also susceptible for its parasitisation. *Cuscuta* also thrives on several weeds that grow on the field bunds and nearly wastelands. The problem of *C. campestris* on alfalfa can be controlled easily through the control of weed *Convolvulus arvensis* as the parasite thrives on weed host. *Cuscuta* can be controlled by dense, well-grown crop of alfalfa because the parasite needs sunlight for its development.

15.8 Using resistant varieties

Older crop plants usually become resistant to the infestation of parasite. Research work is undergoing to develop resistant varieties in pulses and other crops. Varieties that are having high calcium levels are resistant to the parasite. Calcium rich tissues are seems to be inhibitory to enzymes involved in penetration of cuscuta haustoria.

15.9 Chemical control is partially effective

Low doses of glyphosate may provide an alternative to the drastic spot burning if the *cuscuta* is already established in alfalfa. This technique was pioneered in U.S.A. and France. Glyphosate is translocated to and accumulated by the physiologically active growing points of *Cuscuta*. However, *haustoria* embedded in host tissue are not killed and recovery, can occur. Similar technique may also be applicable to other host crops.

Spot spraying of alfalfa with non-selective herbicides like diquat and paraquat can provide as effective as flame throwers. Non-selective chemical burning can be done in crops which they can recover. Some crop plants like alfalfa has underground stems. Dinoseb-acetate and diesel oil also can be sprayed on infected vegetation with success.

Cuscuta establishment can be prevented by spraying certain residual herbicides like chloropropham propyzamide and chlorthal dimethyl in chickpea, chlorthal-dimethyl in carrot and propyzamide in Niger. Spraying of sulfallate plus pebulate in tomatoes control the *cuscuta* effectively.

15.10 Residual life

In tropical countries residual life of herbicides is very less due to rapid degradation. Persistency can be improved by using a mixture of P-Chlorophenyl-N-methyl carbamate or with the carbaryl insecticide which has similar chemistry.

Repeated sprayings of simple calcium salts will protect susceptible plants from attack. In sugar beet, propyzamide applied at a rate of 1.5-2.0 kg/ha at the 4 to 6 leaf stage has been found successful in Italy and Soviet Russia. Ethofumesate applied at a rate of 1 kg/ha preemergence or 0.6 kg/ha early post-emergence was also found successful in controlling the weeds. Cycloate at pre-emergence also used in Yugoslavia.

Controlling of weed with maleic hydrazide is quite interesting. The chemical MH sprayed at the rate of 7 kg/ha plus wetting agent, on well established *Cuscuta campestris* on super beet in July. This kills the vines and decrease the viability of any seed already produced. Although the growth of crop is reduced but sugar yields are increased.

15.11 Natural insect enemies

Cuscuta can be damaged by *Smicronyx noridus*, a gall-forming weevil in India. *S. Jungermannia* is highly effective in controlling *C. campestris*. In China, work is under progress to control the problem with mycoherbicide, *Colletotrichum* sp.

15.12 Broomrapes (Orobanche sp.)

The genus *orobanche* originated in the near East and almost all the agriculturally important species are found in this region. All the *orobanche* species are root parasites. The matured plant produce huge quantities of minute seed (0.25 mm long). The seed germinates and produce a "*Procaulome*" a few millimetres long in response to stimulant substances exuded by the host plant root. Attachment to host root must occur within a few days, otherwise, the seedling may die-off. A haustorium is formed and penetrates into vascular tissues of host particularly xylem vessels. A mass of *nodule* develops about 1 cm diameter due to modification of tissues, before the shoot is emerged. The shoot emerges very rapidly and produce flowers within a few days of germination. The parasite is solely depends on host for both mineral water and photosynthesis because of non-chlorophyllous leaves. Though the damage is less to the host in case of less infestation but the parasite interferes with the ability of host plant in the absorption of mineral water from the soil.

The damage is very severe under drought stress conditions. A single emerged parasite on *Vicia faba* may reduce yield by 8 per cent. If the host plant is infested with 4-5 parasites the yields may goes down by 50 per cent.

15.13 Features of prominent species

15.13.1 *Orobanche crenate* Forskal (Syn. O. *speciosa* P.C.)

- Attains to a height of 50-100 cm.
- Branches not present.
- Flowers are white with purple markings.
- Large flowers with a size of 2 cm across.
- Flowers subtended by one bract only.
- Bracteoles not present.

15.13.2 *O. Cernua* Loeff (Syn. *O. Cumana* wallr)

- Attains to a height of 20-50 cm.
- Branches are not present.
- Flowers are pale with dark blue-purple corolla lobes.
- Single bracts present.
- Bracteoles are not present.

15.13.3 *O. ramosa* L

- It is a dwarf parasite and attains to a height of 10-30 cm.
- Had branches.
- Flowers pale is to bright blue up to 2 cm long, 1 cm across.
- Anthers glabrous.
- Some flowers are whitish.
- Stems are yellowish.
- Flowers are subtended by a bract end two bracteoles.

15.13.4 *O. aegyptiaca Pers*

- Attains to a height of 15-40 cm.
- Flowers pale to bright blue.

- Flowers over 2 cm long and up to 2 cm across.
- Hairy anthers are present.
- Stems are yellowish.

15.14 Range of hosts

Orobanche species have restricted host range when compared with *Cuscuta*. But it has less specificity than *Striga* species. It is a severe problem in crops like faba bean, lentil, potato, peas, carrot, safflower and chick pea. The weed *Ammi majus* belongs to *Umbelliferae* is also attacked seriously. Chick pea escapes its infestation if planted late but early platings are severely affected. Crop failures are very common in Egypt, Turkey, Lebanon, Syria, Jordan, Morocco and other Mediterranean countries due to infestation of *orobanche* species.

Orobanche cernua is mostly attacked on crops belonging to *Solanaceae* family like Brinjal, Tomato and Tobacco etc., but not on sunflower. *Orobanche ramosa* and *O. oegyptica* are most troublesome on plants belonging to families *Solanaceae fabaceae, Cruciferae, asteraceae* and *Chenopodicaceae.*

15.15 Infestations should be prevented in new areas

The minute seed of parasite can be distributed in the fields via soil, water and wind. Hence it is important to remove adult plants from the field before flowering to prevent the build up of seed bank. As the *orobanche* seed is very minute size, the detection, and separation from crop seed is very difficult.

The problem can also be solved by the inclusion of non-susceptible crops in the rotation. The control of *O. cernua* is possible with the aforementioned strategy as the parasite has greater specificity. Secondary hosts should also be considered for the control to prevent proliferation and carry-over of infestation.

15.16 Cultural technology

Rotation of crops is not an easy method and requires several years. Sometimes most profitable crops should give up if they are susceptible. Trap crops which stimulate the *orobanche* seed germination may help to hasten the process. Flax has been found to be an effective trap crop for *O. ramosa*.

O. crenata has a more complex germination ecology and a strict pattern of seasonal behaviour and germinating most abundantly in November and December. It also varies with locations. In Egypt, about 18,000 plants/ha are emerged in chick pea crop planted in October but only 4,000 plants/ha was seen in November planting.

Prolonged flooding has been found successful in tobacco for controlling the weed. High fertilization also suppress the weed growth.

15.17 Solarization technique

A promising technique developed in recent years is the use of solar heating. Solarization involves covering moist soil with clear polyethylene film for a period of several weeks at the hottest time of the year. Solarization for about 5 weeks, successfully control the *O. aegyptica* in carrot fields. Temperature of the top soil reaches more than 55°C and many organisms are also killed. The technique is, however, expensive and only appropriate to high-value crops.

15.18 Using verietal resistance

In some of the crops like sunflower and *Vicia faba* few varieties have been found having defence mechanisms against the parasite. However after a few years, they become susceptible to more virulent forms of parasite. Detailed studies have to be established that the resistance results from a lignification response in the stele of resistant roots. Genetic studies indicating that resistance is associated with single dominant genes.

Resistance is also some what affected by climatic conditions and fails under drier conditions.

15.19 Biological means of control

Biological studies would appear to be overdue in the light of very considerable success. The agramyzid fly *Phytomyza orobanchia* has been found to control *orobanche* sp. in Russia and Eastern Europe. Shoots of infested plants are collected at the end of season, complete with pupae in the seed capsule and stems. They should be stored under suitable conditions (6-8°C and 75-85% humidity) over winter and then hung in fields where *orobanche* is emerging.

The fungus *Fusarium* sp. can also be used to control *O. aegyptica* by inoculation of the planting holes of the host crops tobacco and melons.

15.20 Chemical control

- Spraying of Alkyl alcohol at 0.1 per cent concentration is alternative to hand pulling.
- Oils can be used but there is difficulty in avoiding damage to host crops.
- Spraying of glyphosate (0.2 kg/ha) against *O. crenate* on *Vacia faba* is an effective method.
- Repeated lower doses (60g a.i./ha) of glyphosate is better than higher (1220 g/ha) single dose.
- Herbicides such as trifluralin and diphenamide provide partical control of *Orobanche* sp.
- Seed treatment of *V. faba* with a mixture of fluazifopbutyl in oil is also effective.
- Application of Chlorsulfuron (1-2 g a.i./ha) control *O.crenata* in *V. faba*.
- Fumigants like methyl bromide is also highly effective.
- Methan-sodium and ethylene dibromide is less expensive.

- An alternative chemical approach is the use of substances that trigger suicidal germination in the absence of host.
- The alternative germination stimulant is ethylene gas which can be used successfully.

15.21 Witch weeds (*Striga* sp.)

Witch weeds are mostly distributed in tropical countries rather than temperate world. They are less spread in near East region. However in Sudan they effect much greater areas and cause heavy losses than all other prasitic weed groups. It is also a root parasite and has similar biology with *orobanche* sp. Smallness of the seeds makes them obligate parasites and able to develop fully only after attachment to a suitable host root. The seeds germinate only in response to stimulating root exudates, within a few millimetres of a crop root. After attachment, they develop as total parasite below ground. On emergence into sunlight from soil, the plants can synthesis chlorophyll and photosynthesize and are no longer total parasite. That is why it is called hemiparasite. The efficiency of photosynthesis in *striga* plant is less efficient and is only about 20 per cent of normal plant.

15.22 Side effects

The parasite does continue to draw heavily on the host photosynthates but also causes other serious changes in the host physiology. The infestation results in stimulation of sort systems but serious reduction in shoot growth. The crop losses are so far not estimated properly. Reliable estimates indicates that crop losses may occur up to 25-50 per cent. The losses are severe in case of infertile soils and rainfall is erratic.

15.23 Distinguishing features of species

15.23.1 *Striga hermonthica* (Del) Benth

- Attains to a height of 15-100 cm.

- Flowers are pinkish coloured.
- Had five calyx ribs.

15.23.2 *S. asiatica* (L) Kutze (Syn. *S. lutea* Lour)

- Attains to a height of 15-30 cm.
- Flowers are white, sometimes drying brown.
- In some countries like Arabian peninsula the flowers are scarlet or yellow coloured.
- Had ten calyx ribs.

15.23.3 *S. densiflora* Benth

- Reaches to a height of 15-30 cm.
- Flowers are white and somewhat drying bluish.
- Had 5 calyx ribs.

15.23.4 *S. angustifolia* Saldanha (Syn. *S. emphrasioides*)

- Attains to a height of 15-40 cm.
- Whitish flowers.
- Had 15 Calyx ribs.

15.24 Host specificity

Striga species have varying degrees of host specificity. *S. asiatica* occurs more widely in Sudan but not attack field crops. *S. hermonthica* attacks most cereal crops, especially sorghum in Arabian peninsula and Pakistan. It also attacks *Pennisetum* millet, upland rice, maize, finger millet and sugarcane. A number of races in *S. gesneriodes* affects cowpea in West Africa. It also attacks *convolvulaceae* and *Euphorbiaceae* in Arabian Peninsula. *S. densiflora* infests cereal crops and sugarcane in Pakistan. Some wild grasses, cereals and sugarcane are attacked by *S. angustifolia* in Pakistan. *S. latericea* distributed in Somalia and Ethiopia and problematic in sugarcane.

15.25 Prevent the spread of parasite

The introduction and spread of *striga* is more difficult to prevent than that of *cuscuta*. As the seeds are minute (0.25 mm long) it can readily adhere to crop seed. It can be blown by wind and carried by water over long distances. The crop seed should be obtained from uninfested areas only.

Another form of prevention of parasite is to avoid growing susceptible crops. Unfortunately, majority crops in drought prone areas are highly susceptible to the parasite. Except cowpea, several crops are resistant to *S. gesnerioides*. There is a strong specificity of some populations of *S. hermonthica* and *S. asiatica* to either sorghum and some other millets.

15.26 Host specificity

Striga species have varying degrees of host specificity. *S. asiatica* occurs more widely in Sudan but not attack field crops. *S. hermonthica* attacks most cereal crops, especially sorghum in Arabian Peninsula and Pakistan. It also attacks *Pennisetum* millet, upland rice, maize, finger millet and sugarcane. A number of races in *S. gerneriodes* affects cowpea in West Africa. It also attacks *convolvulaceae* and *Euphorbiaceae* in Arabian Peninsula. *S. densiflora* infests cereal crops and sugarcane in Pakistan. Some wild grasses, cereals and sugarcane are attacked by *S. angustifolia* in Pakistan. *S. latericea* distributed in Somalia and Ethiopia and problematic in sugarcane.

15.27 Prevent the spread of parasite

The introduction and spread of striga is more difficult to prevent than that of *cuscuta*. As the seeds are minute (0.25 mm long) it can readily adhere to crop seed. It can be blown by wind and carried by water over long distances. The crop seed should be obtained from infested areas only.

Another form of prevention of parasite is to avoid growing susceptible crops. Unfortunately, majority of crops is drought prone areas are highly susceptible to the parasite. Except cowpea several crops are resistant to *S. gernerioides*. There is a strong specificity of some populations of *S. hermonthica* and *S. asiatica* to either sorghum and some other millets.

15.28 Cultural means

The land should be rotated with non-susceptible crops to avoid the parasitic problem. This is also solution for reducing the problem on a long run. Surprizingly, the longevity and viability of seeds in the soil may be atleast 5-10 years. Severely infested fields may require growing of non-susceptible crops for several years.

15.29 Using trap crops

Certain non-host crops acts as *'trap crops'* exuding a germination stimulant from their roots. It causes *'suicidal'* germination. These crops appear to be beneficial for exhausting the seed from germination biology of *striga* species. The factors which are responsible for the seed germination should be studied more deeply.

The preference of striga under dry conditions is not fully understood. Sometimes even copious irrigation cannot suppress the germination. Planting should be done under continuous rainy period for less infestation. Early planting followed by dry spell encourage the parasite seed germination.

15.30 Flowerishes on submarginal soils

Low soil fertility is also associated with high striga infestation in several crops. Application of N-fertilizer is beneficial to the crop and delay the striga emergence. Nitrogen helps in maintaining high shoot, root ratio in the host, which unfavour the parasite. The fertilizer also creates greater shading of soil due to high leaf area index. Lowering soil temperature also reduce the germination of seed in the soil. An optinum temperature of 30° is needed for the seed germination. Increasing crop density is also one way to suppress the parasite.

15.31 Continuous weeding

Striga usually emerges very much later than majority of weed flora. It starts infestation 4-12 weeks after sowing depending upon rainfall prolongation. This late emergence makes it avoidance of main weeding operations. Continuous weeding is necessary to prevent the building up of the pest. However, prolonged weeding is regarded as a very difficult task and is needed to prevent the germation of enormous minute seed. Leaving the field without weeding leads to maturation of thousands of weed seed and development of seed bank. Continuous hand pulling and hoeing is quite genuine during longer growing seasons to apply continuous pressure on the weed. The operation is to be repeated once in every two weeks to arrest the seed production. Prevention of seed production is a vital component of any long-term control programme of parasitic weeds.

15.32 Hand pull of mature plants

Hoeing young plants is not desirable because of their regrowth from many laterals below the ground. The most ideal thing is to pull the mature plants just before flowering, when their tougher stems will ensure that in the whole plant is removed. Where infestations are too heavy for hand pulling to be practicable, herbicides may provide an alternative. The infestations should be delayed by using techniques like dense crop populations nitrogen fertilization and growing resistant crops or varieties.

15.33 Selection of resistant varieties

Several varieties of sorghum resistant to striga are screened at International Crops Research Institute for Semi-Arid Tropics (ICRISAT) and made progress in this time. New varieties like SAR-1 and SAR-2 are found to be promsuing for *Striga asiatica*. The variety SRN-4841 was found to be resistant to *S. hermonthica*. But these red grained types are not preferred by the consumers. White grained derivatives are recently developed in some regions.

Semi-wild *Shibra pennisetum* millet is also showed resistance to striga. In cowpea, Suvita-2 and Suvita 58-57 varieties are highly resistant to *S. gesneriodes*. Most of the resistant varieties have low exudation of germination stimulants from their roots. Failure of parasite attachment and development of host due to mechanical barriers or chemical factors are not yet known.

15.34 Chemical control

Grown up striga plants can be killed readily by spraying herbicides directly on the shoots. Usually paraquat and 2, 4-D are used in maize fields whereas bromoxynil and ametryne are used in case of legumes. These chemicals not only stop the seed production of parasite and decrease the crop losses. The timing of application in relation to germination and attachment of the parasite may be critical. Rainfall is also required to leach 2, 4-D herbicide into the soil. However, 2, 4-D is not provide cent per cent control. Sometimes oxyfluorfen is used for prevention of striga emergence. The shoot tips of the weed are killed when they reach the soil surface. *Striga hermonthica* which infested on sorghum can be effectively controlled by potent sulphonylurea chemical chlorsufuron at the rate of 1-4 kg/ha. Metylbromide, a volatile fumingant is used for localized control. An interesting method to stimulate the seed germination by using the ethylene gas is most practicable. The gas may be injected at the 1-2 kg/ha at a depth of 15-30 cm which result in 80-90% seed germination of parasite. This stimulation of germinations is *'suicidal'* to the parasite.

15.35 Biological means

Striga can be controlled through some insects. An introduction of moth *Eulocastra argentisparsa* to Africa from India has given results. Similarly gall froming weevils *Smicronyx* spp are introduced into India from Africa to control the weed with moderate control.

Recently, interest is increasing for the development of '*Mycoherbicide*' by inundation of fields with spore suspensions of indigenous fungi for the control of striga.

15.36 Mistletoes (*Viscaceae* and *Loranthaceae*)

The mistletoes belongs to both families creates problems in fruit trees and ornamental trees. Growth of mistletoes results in death of the affected branch beyond the point of attachment. It also make the tree more susceptible to drought stress. Usually some birds carry the seeds from tree to and spread the infestation. Germination and establishment of seedings is slow and populations take few months to build up serious infestations.

15.37 Cistanche spp (*Orobanchaceae*)

These are wide spread in mediterranean region. In desert regions they attack wild vegetation especially perennial *Chenopodiaceae*. They are very handsome, beautiful yellow flowered and grow to a height of 1 m. Sugarbeet is also another host for this parasite.

15.38 *Thesium humile* vall

It is a root parasite in the family *Santalaceae* has been reported as a problem in cereals and onion.

15.39 *Melampyrum arvense*

It is a facultative hemiparasite of the *Scrophulariaceae* family occurs as problem in cereals. It can successfully controlled by MCPA and bromoxynil.

16

SOIL SOLARIZATION

Make hay while the sun shines
Strike when the iron is hot

—*English proverb*

Soil solarization is a technique of increasing the temperature of soil with the help of solar radiation. It involves mulching of the soil with clear plastic films so as to trap the sun's heat in the surface soil. Hitherto chemicals are used to control soil pests including weeds. Use of chemicals for pest control is hazardous to the environment and costly. The new method of solarization is a novel technique of controlling soil borne pests including weeds. The enhancement of temperature due to hot sun would be lethal to soil inhabiting macrofauna, nematodes, pathogens and weed seeds.

16.1 Historical background

The idea of soil solarization was based on observation by farmers in the Jordan valley who observed the intensive heating of the polyethylene (PE) mulched soil. Initially some of the Israel researchers have developed disinfestation of soil through solar heating successfully. Since then the technique has attracted the attention of several researchers from all over the world. They have been noticed the importance of solarization in controlling soil borne pests.

16.2 Benefits of soil solarization

- It is an unique non-hazardous technique.
- Environmentally benign.
- User-friendly and environmentally safe.
- It is not dependent on fossil fuel or any costly chemical.
- Effective on broadspectrum of microfauna.
- Kills pests like soil borne nematodes, fungi, bacteria and weed propagules.
- Effective for one season to one year in controlling weeds.
- Solar radiation is an infinite source of energy and supplied by nature on free of cost.
- It helps in mineralization of soil micelles.
- It stimulate crop growth by increasing nutrient availability.
- The technique is simple and easy to use by farmers.
- It also enhance pesticide degradation in soil.
- Savings in nutrient application.

16.3 Technique of solarization

Soil solarization involves covering the wetsoil with thin transparent polyethylene films during the hot summer months. This process would raise the surface soil temperature by 8-12°C as compared to non-solarized soils. A duration of 4-8 weeks is sufficient to give satisfactory control of weeds. Several annuals, parasitic weeds and some perennials are sensitive to solarization. Some perennials like *Cyperus, rotundus, Conyza canadensis, Cynodon dactylon* and *Melilotus indica* are not controlled by solarization.

16.4 Characteristics of Polyethylene film

The purpose of solarization is trapping of maximum solar radiation through transparent films. The characteristics of PE film for efficient harnessing of sun's radiation is given hereunder:

- The films should be transparent for easy penetration of sun's energy into the soil.
- Black films are not clear transparent and hence not useful.
- The film should have high strength.
- The films should be thin (19-25 μm) for effective solar heating.
- It should be less expensive and highly durable.
- The film should reduce heat loss from soil.
- The film should prevent evaporation and heat conversion from soil.

16.5 Factors that influence solarization

A few factors influence the effectivity of solarization technique. They are : soil character, soil moisture, thickness of PE film, type of weed species, present in the soil, time or period of exposure and season.

16.6 Soil colour

Black coloured or dark soils absorb radiation quickly without any reflection losses than light coloured soils. Heavy textured soil absorb the heat slowly but retain longer than light textured soils. Good land preparation helps in absorption of more heat than compact and uneven soil.

16.7 Soil moisture

Moist soil usually transfers heat quickly than dry soil. Irrigation before mulching make the soil thermal sensitive. The soil water helps in transfer and conduction of heat in the soil body. Saturated condition is optimal. The microflora and fauna is activated with moisture and susceptible to thermal killing.

16.8 Thickness of polyethylene film

Thinner films (19-25 μm) are more effective for solar heating than thicker films (50-100 μm). Maximum temperatures in upper soil layers under ideal conditions are achieved within 3-4 days after solarization started. The upper 15–30 cm of soil show diurnal temperature changes influenced by day and night atmospheric temperatures. Soil temperatures will be increased about 8–10°C than corresponding non-solarized plots. The temperature increase was 9–7°C with 100 μm film whereas it was 13.8°C with 50 μm film. Batter effect was also reported with double layer of films.

16.9 Season suitable for solarization

The solarization should be practised in warmest summer months. In tropical countries, a duration of 4–6 weeks would be adequate in several situations. In *South India*, April and May months and in *North India* April May, June are best months for soil heating. Heating of soil is best when plastic film is laid close to the soil with a minimum of air space to reduce the insulating effect of air layer. Usually solarization effect is restricted to surface soil only (0-10 cm depth). Solarization is reported to reduce seedling numbers to less than 10 per cent.

16.10 Susceptible weed species

Elevated temperatures due to soil solarization results in reduction of weed emergence. All weeds are not susceptible to solar heating due to some adoptations. ***Winter weeds*** are highly susceptible to solar heating.

The following annual weeds are controlled by solar heatings :

Ageratum conyzoides	*Digera arvensis*
Achrachne racemose	*Digitaris horizontalis*
Abutilon theophrasti	*Echinochloa colona*
Amaranthus retroflexus	*Euphorbia hirta*
Amaranthus spinosus	*Eragrostis japonicum*

Amaranthus viridis
Anoda cristata
Anagallis coerulea
Avena fatua
Avena ludoviciana
Avena sterlis
Brachiaria mutica
Bidens pilosa
Chenopodium album
Chenopodium murale
Commelina communis
Cynodon dactylon
Dactyloctenium aegypticum
Hypericum crispum
Lamium amplexicaule
Orobanche aegyptica
Oscimum sanctum
Phalaris minor
Portulaca oleracea
Rumex dentatus
Sida spinosa
Sonchus oleraceus
Sorghum helepense
Trianthenum portulacastrum
Tribulus terrestris
Xanthum spinosum

16.11 Sensitive to solar heat

The dominant rainy season weeds like *Trianthema monogyna, Digera arvensis, Dactyloctenium aegypticum,* and *Echinochloa colonum* and winter weeds such as *Avena ludoviciana, Phalaris minor, Chenopodium album, Rumex dentatus* and *Fumaria indica* are highly sensitive to solar heating and readily killed by solarization. The seed-borne sedges also highly susceptible to the treatment. However, *Cyperus rotundus* tubers are tolerant and survive even in prolonged solarization. This is because of its presence in deeper layers. The effect of solarization diminishes with soil depth. Weeds which are capable of putting up growth in deeper layers survive for solar treatment. The winter weed *Melilotus* spp. have high heat tolerance due to its thick seed coat. There was an increase in population of this species even at 40 day solarization.

Excellent control of *orobanche* spp. in host plants were observed in Israel and Syria due to solarization. This is quite advantageous where tobacco fields are severely infested with broom rape. Solarization for the period of 40 days with a transparent polyethylene film of 50 μm thickness controls about

60 per cent *orobanche* parasites. Besides, about 70 per cent of dicot and monocot weeds are also controlled effectively. Variabilility of weed control may be influenced due to size and depth of seeds, vegetative propagules and soil type.

16.12 Drawback of the system

Soil solarization, undoubtedly controls majority of harmful fungi, nematodes, bacteria, insects and weeds particularly those propagating through seeds. However, its effectivity is limited to surface soil. Weed seeds present in deeper layers of soil were not affected. The weed seeds which are brought up to the surface by any means creates problems as usual. If the land is cultivated the weed seeds coming up to surface and nullify the solarization effect.

16.13 Deep ploughing may be avoided

Direct seeding without any soil disturbance is desirable to avoid deep seated weed seeds. The solar effect is best in crops which form quick canopy cover. The escaped weed seeds from solarization may compete with crop plants if the slow growing crops are sown. The population density of weeds markedly decline in treated plots. Manual weeding, low cost chemical weeding and other low-energy inputs would prove highly cost-effective and proved to be beneficial.

16.14 Buried weed seed Banks

The seed reserves of dormant weeds acts as a reserve bank and creates prolonged weed problems in cultivated fields. Repeated control measures in an integrated manner gradually reduce the number of dormant seeds. Soil solarization would help for eliminating the wed persistency to a considerable extent. However solarization does not eliminate strongly dormant weed seed from the germination zone. Solarization kills most of non-dormant seeds and greatly decrease the weed seedlings in soil. Weed seeds present in upper 5 cm depth of soil will be killed by solar heat. However, its effect is poor on weed seed present in deeper layers where the solar heat cannot diffuse. Some susceptible weeds like *Commelina communis* and *C. benghalensis* can be controlled upto 10 cm depth with 40 days of solarization. Some grassy weeds like *Echinochloa crus-galli* can be controlled up to 4 cm of soil depth with the solar treatment. In deeper layers, usually bold and coarse seeds are able to emerge from the soil but tiny seeds lie dormant for several years until it comes to the surface zone.

16.15 Tolerance of certain species

Some weed species are highly tolerant to solarization because of their genetic make up, thickness of seed coat and enzyme mechanisms.

16.15.1 *Weeds tolerable to solarization treatment*

Amaranthus spinosus
Astragalus sp.
Conyza canadensis
Cynodon dactylon
Cyperus esculentus
Cyperus ratundus
Digitaria marginata
Desmodium spp
Melilotus indica
Melilotus sulcatus
Malva speciosa

Several perennials are capable of putting up growth in deeper layers and solarization effect was nil on these weeds.

16.16 Nutrient transformation

Extensive studies on plastic mulches indicates that significant changes occurred in nutrient transformation of soils during solarization. Some nutrients such as ammoium-N, nitrate-N, Calcium and Magnesium are markedly enhanced in the soil. The availability of micronutrients such as Fe^{+2}, Mn^{+1}, Zn^{+2} and Cu^{+2} also augmented due to thermal activity. Electrical conductivity of soil is also increase due to solarization. However the organic carbon of soil is not altered. Extensive studies in several countries indicates that increase in nutrient availability is transient but not persist long. The humus content of soil is rapidly decomposed and release nutrients due to raise in soil temperatures.

16.17 Change in the activity of soil microflora and fauna

Soil is not a heap of inert mass but also the seat of innumerable forms of plant and animal life. Among *micro-fauna* millipeds, spring tails, nematodes, snails and slugs are mostly found in surface layers of soil. *Microflora* such as protozoa (Rhizopods, flagellates and ciliates), Fungi, actinomycetes and bacteria also function actively in superfluous layers where organic substrates and moisture is present and play an important role in nutrient mineralization.

Partial sterilization of soil through burning stubbles was known to Romans in ancient days. Similar fact has long been known to Indian farmers. Heating the soil either through organic debris or scorching sun has been practised from time immemorial in India and Egypt. Heating the soil has been believed to improve in soil fertility. Exposure of soil to the shining sun is an ancient practice still common in South India and Egypt.

Presence of protozoa along with other organisms retards the bacterial activity and when the partial sterilization of soil was done they were destroyed resulting in indirect increase in number of beneficial bacteria. Exposing the soil to scorching sun heat eliminates undesirable pathogenic organisms from the soil.

Shift in the microbiota of the soil following solarization is in favour of antagonists. Weed seeds are weakened and weed seedlings are killed due to solarization.

16.18 Native rhizobia

Solarization, in general, decrease the native rhizobial population of pigeonpea and chickpea crops. However, the treatment do not alter rhizobial population in unirrigated soils.

Investigation carried out at Israel, is sufficient to cause heavy nodulation of bean roots survived after solarization. Increased colonization of plant roots by plant growth promoting *fluorescent pseudomonas* from inoculated seed also occurred following soil solarization.

16.19 Increased growth rate (IRG) of crop plants

Soil solarization has tremendous effect on soil-borne pathogens, weeds, harmful protozoa, fungi and eelworms. The treatment enables the crop plants to grow vigorously and yield better as compared to non-solarized plots. There are many reports of enhanced growth and yield of crops. The magnitude of increase depends upon the nature of pest, type of soil, and degree of control. Checking of weeds due to solarization technique enhanced the yield of soyabean by 78 per cent, sesamum by 72 per cent, groundnut by 52 per cent and onion by 100-125 per cent. It is also highly effective in eliminating parasitic weed *orobanche* and an yield of 78 t/ha of carrot was reported from sun baked plots, while the unsolarized plots yield nothing. Similarly bean yields also enhanced by 20 per cent due to control of *orobanche*.

Solarization is also enhances the crop yields in fields where no infestation of weeds and nematodes because of some chemical and biological changes occurred in the soil system.

16.20 Practical feasibility

Solarization technique alone, could not enhance growth and yield of crops. It is an integrated method of increasing plant health and yield. Climatically Indian subcontinent is more suitable for solarization and almost every area of cultivated field can be treated with solar radiation. The country experiences mean maximum temperatures of more than 40°C in the hot summer months (April to June). During the summer, all cropped lands are kept under fallow enabling farmers to practice solar treatment without sacrifycing his crop.

16.21 Economics of practice

The major constraint is the high cost of polyethelene films in India. Thinner films are effective and also less costly. The period of use of polyethene sheet is limited to only one and half months and same material can be use once again in the next year. While computing the economic control of pests, savings in nutrient application and enhanced yields, have to be taken into consideration. The solarization treatment also helps in degradation of pesticide residues in soil rapidly and decontaminate polluted soil.

17

MANAGEMENT OF NOXIOUS WEEDS

Evil weeds and enemies should be nipped in the bud.

—French proverb

17.1 Wild sage (*Lantana camara*)

The wild sage is very commonly called as *Lantana* or Australian berri in India. However, it is also known by *Punch phul buti, Churel buti, Bunch berri, Red phulano, Jackal grape* etc. It is a perennial shrub belonging to *Verbenaceas* family, grown luxuriously in moderate to heavy rainfall areas. It is usually seen in non-cropped areas through- out the globe. The plant vary from compact to spreaded shrub and grows up to 5 m height under favourable conditions. Even under adverse conditions it attains to a height of less than one metre. The size of weed varies with agroclimatic conditions of the region. The flowers are small and exhibit variety of colours ranging from scarlet, pink crimson, yellow and white.

17.1.1 *Wide distribution*

It is well distributed in all tropical countries, although its native place is Tropical America and West Indies. It was introduced in India from Australia as ornamental plant in 1809. Its spread is like torrential flood in U.P., Bihar, M.P., Jammu

& Kashmir, Rajasthan, Punjab, Haryana, Himachal Pradesh, Maharashtra, and Kerala. Thick stands of weed is also found in Deccan plateau and Deltaic regions of eastern coast.

17.1.2 *Phenology*

Vegetative growth is very active during the months from March to June when high temperatures prevails. It produce flowers and fruits during June to October when the rainfall is high. It lies in dormant state during November to March when low temperature exists.

The weed has long tap root system with prominant lateral branches and difficult to pull by manual labour. The stem is pubescent and emits some pungent odour. The leaves are thick, deep green, crenate and short stalked. The leaves emits phenolic odour when rubbed. The flowers are borne in small axillary heads. The fruits are glossy green when young but turn to purplish black on maturity.

17.1.3 *Animal poison*

Stray cattle grazed on the weed usually die within one day. Several deaths has been recorded in Himachal Pradesh, Uttar Pradesh and Maharashtra where animals feed on lantana infested pasture lands.

Lantadena A. has been considered as most important toxic constituent in this weed. The fruits of Lantana contain mainly *Lantadene* which is responsible for some nervous and heart ailments in children when eaten. The green leaves contain high quantities of *Lancamarene* which is responsible for animal deaths.

17.1.4 *Possible utility*

In rural areas, Lantana bushes are cut and dried and used as a fuel wood. Heavy smoke and bad odour are the drawbacks of this fuel. The twigs are used for manufacturing furniture and vegetable baskets. The stalks of Lantana can be used as raw material. Lantana dry mass can be used as a mulch in row crops and in orchards for moisture conservation and to suppress weedflora. The decomposed litter enrich the surface soil.

The greenleaf twigs of Lantana are incorporated in puddled soils as green leaf manure in some areas. Research indicates that it helps in saving of 40 kg N/ha and enhanced rice yields. This practice also suppress some of the *Cyperaceous* and *gramineous* weeds arise from rice fields. Lantana biomass can be used for the generation of biogas through anaerobic fermentation.

Soil and its fertility can be restored when it grown on denuded rocky hills. Land sliding and soil erosion can be arrested when lantana is grown in steep slopes.

17.1.5 *Physical means*

Stumping of plants up to 25 cm above the ground and burnt the left over stumps after some time gives good control. The stumps can be pulled out with crowbars during summer. The regrowths can be controlled with frequent cutting and burning for about 3-4 years continuously. Slashing, burning and uprooting requires large number of manual labour.

17.1.6 *Cultural manipulation*

Wastelands should not be kept vacant. Growing castor (*Ricinus communis* L) rubber (*Ficus elastica*), Mexican sunflower (*Tithonia diversifolia*), and *Puperaria thumbgiana* are effective for arresting the *Lantana* infestation. Planting perennial pigeon pea (*Cajanus cajana*), fodder trees, and fodder grasses like napier, guinea and setaria grasses checks the spread of the weed.

17.1.7 *Elusive biocontrol*

Government of India explored the possibility of using tingid bug (*Teleonemia scrupulosa*) against the weed during 1916 with little success.

This insect was imported from Australia but found unsuccessful on large scale because the insect remains in epidemic proportions in certain seasons only.

17.1.8 *Chemical ways*

The Lantana bushes should be cut down during premonsoon season up to a height of 5-10 cm above ground. After one month, enough regenerated foliage will develop from cutted portions during rainy season. This foliage should be wetted with one per cent solution of glyphosate (glycel 41SL) with the help of hand sprayer. This will kill the plants completely.

17.2 Congress grass *(Parthenium hysterophorus)*

The congress grass is exotic weed and a native of Central and North America. It has been introduced into India along with imported wheats by an accident. It first appeared in Mumbai and Pune in 1956 and rapidly spread in peninsular India during sixties. This noxious weed is started fast spreading during 70s in entire North India. It has been observed that irrigation water played a major role in spreading the weed in crop fields. Floods during rainy season also helped the weed for rapid spreading in non-cropped areas. The weed has spread as a ruderal throughout Indian subcontinent by 1987 and occupies more than six lakh hectares in both cropped and non-cropped areas in mainly due to its high competitive ability, suppression of neighbouring plants, high ecological amplitude and allelopathic effect.

17.2.1 *Morphology and phenology of weed*

The weed belongs to family *Asteraceae* and probably has been derived through hybridization of two species, *Parthenium confertum* and *P. bipinnatifidum*. The plant attains to a height of 1.0-1.5 m and profusely branched. This annual herb is having chrysanthemum like leaves or irregularly dissected carrot like leaves and stem is covered with hairs. The stem is longitudinally grooved. The seeds are ovate or ellipsoid with remmant of corolla and style.

The congress grass is well known as *"Carrot weed" "White top", vayyari bhama, "Gazar ghas", "Chatak chandam", "Gandhi booti"* and *"Macherla Kampa"*, is threat to agriculture and hazardous to human and cattle health.

The weed complete the life cycle within 120 days. The emergence and periodicity, mainly depends upon frequency and amount of rainfall during the year. The plant completes 2-3 generations within one year. It is a rapid colonizer and completes with other vegetation successfully. Phytosociologically, the weed is highly aggressive and suppresses the surrounding vegetatum through competition, allelopathy and amensalism.

17.2.2 *Movement of weed*

The seeds are easily blown by speedy winds to distant places. The seed also carried away by rain water, irrigation channels and drainage canals to lowlying areas. The weed moves to agricultural lands through composts, farmyard manure, city wastes, sewage and irrigation waters.

17.2.3 *Habitats*

The seeds germinate over a wide range of temperatures, but maximum germination occurs at 25-30°C. It can also germinate in continuous light and dark regimes. The weed prefers shady, moist and rich organic habitat and adopt very well to climatic extremes. The weed is abundantly found in wastelands, industrial sites, railway tracks, petrol pumps, railway backyards, bus stands and road sides in urban areas. It is also sighted in grazing lands, orchards, canal bunds, and along highways in rural areas.

17.2.4 *Hazardous nature*

Congress weed has drawn much public attention because of its hazardous nature. Labourers who frequently contact with the plant during field operations caused allergy, eczema, pollentonsis, asthama, gangerence and dermatitis. The sensitive persons are susceptible even air borne dust of trichomes present on the leaves and stems. The dricd plant material is also harmful to humans. The children before puberty are spared but semiclad individuals are prone to all effects of the weed. Usually grazing animal do not touch the plant due to its aroma, disagreeable taste and presence of trichomes. Sometimes hungry cattle graze on the weed wilfully or accidentally for want of green fodder during hot summers. This may cause serious disease syndrome to milch cattle. The cattle grazing in parthenium infested area gives less milk and the milk contains a toxin called "*parthenin*".

Various parts of plant yield phenolic acids which are water soluble growth inhibitors. These phenolics reduce the growth of associated crops to an extent of 50 per cent. N-fixing rhizobia present in legumes are reduced because it suppress colonization of bacteria and reduce the nodulation to an extent of about 60 per cent.

17.2.5 *Mechanical means*

The infested areas of wastelands, gardens, orchards and roadsides should be ploughed thoroughly before flower setting. Mouving should be repeated in lawns, playgrounds and wastelands provide an effective control. During the rainy season the plants may be cutted with sharp sickles or hand tools with a long handle will below the ground level by avoiding contact with the plant. The cutted weeds may be heaped at one place and burnt with paddy straw and kerosene to destroy the seed. This operation should be on community level. Such repeated, campaigns in infested areas make the weed under control.

7.2.6 *Biological suppression*

Some aggressive broadleaved weeds like *Casia tora* and *Casia sericea* can suppress the congress weed. In some parts of Karnataka and Maharashtra States the plant is used successfully for control of weed. The *Casia* spp. can be grown vigrously where the rainfall is more than 85 cm. The seeds will multiply themselves and there is no need of fresh sowings every year. The leachates

"Kolines" accumulates in soil when *casia* is grown in the area, inhibit the growth of parthenium. Cattle do not graze on *Casia* spp and being legume it fixes nitrogen in the soil.

The leaf eating beetle, *Zygogramma biocolorata* pallister, highly specific on parthenium. It multiplies on parthenium only. The insect is active during rainy season and attacks the terminal and axillary buds. Later, the leaves are destroyed by the insect. These insects are capable of lying in dormant over cold and hot seasons by burrowing into the soil. The insects remain in sleepy stage for about 24-32 weeks in the soil and emerge after the onset of monsoon. Hence there is no need of reintroduction of beetles every year. The vegetable mite (*Tetranycus cucurbita*) also infest the weed.

17.2.7 *Chemical elimination*

In non-cropped areas, spraying with 2, 4-D amine salt (2-4 kg/ha), or picloram (10 kg/ha) or glyphosate (2-2.25 kg/ha) or Atrazine (1.25 kg/ha or Urea (20% solution) or sodium chloride (15 %) solution on the plant was quite effective at all stages of growth.

In cropped areas, preemergence application of atrazine (1.5-3.0 kg/ha) in maize, pea and other millet controls the weed upto 5-8 months. Application of 2, 4-D (-1.5 kg/ha) as post emergence control the weed in sorghum and maize fields.

17.3 Wood sorrel *(Oxalis latifolia)*

The wood sorrel family (*Oxalidaceae*) encompasses nearly 800 *oxalis* sp. These are widely distributed over temperate and subtropical regions of the world. Six species of the following are found to be problematic in Indian subcontinent.

Oxalis latifolia	*Oxalis tetraphylla*
Oxalis corymbosa	*Oxalis corniculata*
Oxalis pescaprae	*Oxalis stricta*

Most of the species are distributed in sub-mountain regions of Himalayan ranges. Among all species *Oxalis latifolia* is most troublesome weed in India. It is also known as *khattimithi, khatibuti, phipphuru* and *tipatia* in various linguistic parts of the country. The original home of *Oxalis* spp. is central and equatorial South America.

17.3.1 *Morphology and distribution*

Oxalis latifolia is most problematic weed in orchards and upland field crops during rainy season. This perennial herb has no true stem but having scaly root-stock known as bulb. It is having trifoliate leaves with long stalks. The size of leaves is 2.5-5.5 cm across. The weed comes to flowering during July and set seeds during August.

Oxalis corniculata is mostly seen in Himachal Pradesh. It predominates in Kangra valley, Kulu valley, Simla and Kinnaur districts. The weed flowers during May to September and fruiting takes place from June to October.

It propagated through underground stem bearing bulbs also. The bulbs formed at a depth of 30 cm which helps in perennation of weed. The fleshy leaves contain enough stored food. The bulbs of *O. latifolia* goes to dormany during cold season and sprout in spring season when temperatures raises. The bulbs of *O. pes-caprae* remain dormant during hot weather and sprout in autumn.

17.3.2 *Means of Control*

17.3.2.1*Cultural means*

Repeated deep ploughing during summer and regular cutting with hoes and harrows kills the bulbs of plant. It is having high regenerative capacity and persistence behaviour of seeds and bulbs make the plant difficult to control fodder crop establishment and covering the soil with straw mulches also suppress the weed. Maize stover usually used for mulching purpose. Burning and soil steaming at 85°C is also highly effective.

17.3.2.2 *Chemical means*

Application of Diuron (5.0 kg/ha) in maize, Glyphosate (1.0 kg/ha) in citrus orchands, Sulphuric acid (0.5 kg/ha) in groundnut, Oxadiazon (0.75 kg/ha) in citrus, EPTC (0.5 kg/hae) effectively control the weeds.

17.3.2.3 *Biological means*

In Australia, certain insects feed on the plants and control them satisfactorily in some seasons. The distribution of the weed sometimes was done by mole rats. The bulbs of *O. cermia* was found in cavities and underground tunnels of rats which indicates their spread through rodents.

17.4 Bill goat weed *(Ageratum conyzoides L)*

Ageratum is popularly known as bill goat weed an exotic weed that grows luxuriantly in tropical and subtropical regions of the world. It is also called as *white weed* and *neela phulnoo* is various parts of India. This *Ageratum* genera has two species i.e. *A.geratum conyzoides* and *A. houstonianum*. Both species are originated from tropical America and belongs to *Asteraceae* family. The former is an annual and latter is a perennial. Both species are reported to be poisonous to grazing animals. The contact and eaten of the weed causes allergy in humans and grazing cattle.

17.4.1 *Habit and Habitats of both species*

The annual *Ageratum* (*A. conyzoides*) colonizes both cultivated and wastelands. The perennial *Ageratum* (*A. houstonianum*) found mostly in uncultivated lands, orchards, forests canal banks, grazing pasture lands, slopes of natural springs, water channels, and sides of railway tracks and roadways.

17.4.2 *Some of the differences between two Ageratum species:*

Trait	*A. houstonianum*	*A. conyzoides*
Branching	Decussate	Opposite (4-6 plant)
Inflorescence	Capitulum	Capitulum
Heads/plant	80-100	50-75
Flower colour	Sky blue when young white on maturity	Fixed sky blue
Plant type	Robust	slender
Leaves	Ovate triangular	Broadly ovate or ovate
Habit	Perennial	Annual
Habitat	Wastelands	Cultivated/uncultivated lands
Origin	Tropical America	Equitorial America

Annual *ageratum* thrives well in both arable and wastelands in rainy and winter seasons. The infestation is very heavy (800-1000 plants sq. m) in upland crops during rainy season. In winter though infestation is found in all crops but degree of infestation is somewhat low. It completes it's life cycle within 12 weeks. If the weed is uncontrolled properly, the infestation is severe in the following seasons. Its reproductive capacity is tremendous. Each plant produces about 95,000/plant.

17.4.3 *Possible utility*

The flowers are used as ornamental and source of pollen for honey bees. The leaves are used as a vulnery and prevent tetanus. The fresh juice is reported to be a good remedy for prelapsusani. The juice is boiled in sesamum oil and applied in rheumatism. The herbal infusion is given in stomach ailments like dysentry, diarrhoea, intestinal colic and flatulence. It is also good medicine for some type of levers.

The *perennial ageratum* contains an essential oil with agreatochromene, friedlin, friedlin 38-01, stigmastered, potassium chloriop and B-sitosterol.

The weed along with cowdung is also used for production of biogas in rural areas.

17.4.4 *Harmful effect*

The perennial *Ageratum* serve as a possible alternate host of *Pseudomonas solanacearum* a casual organism of wilt disease. The *Ageratum* causes irritation of eye, skin and giddiness in human beings. It also causes health hazard to grazing animals.

174.5 *Unfeasible manual removal*

The root system of *Ageratum* is not so strong and can be uprooted with manual labour. The operation requires tremondous labour force which is not possible in labour scarcity areas. Removals by hand also not possible due to strong foul smell which results in vomiting sensation and giddiness to operators. The weed can be burned with flame throughers. Repeated removals before flowering is an effective method but it should be adopted on community basis. If the cultivated fields are heavily infested in maize-wheat rotation, maize may be avoided in rotation. If the wetland rice is cultivated in one or two years, the infestation is reduced drastically to a minimum.

17.4.6 *Chemical means*

In crops like maize, use of atrazine (3-5 kg/ha) at pre-emergence given good control. Similarly alachlor (1-2 kg/ha) or Oxyflourfen (0.25 kg/ha) can be used in pulses. In wheat fields, isoproturon (1.25 kg/ha) and 2, 4-D (0.8 kg/ha) can be effectively used.

In non-cultivated fields glyphosate (1.0 kg/ha) paraquat (0.4 kg/ha), Diuron (3-4 kg/ha) and common salt (20% solution) can be used with complete success.

17.5 Crofton *(Eupatorium adenophorum)*

The crofton is locally called as *'Kali basanti'* and belongs to the family *Asteraceae*. Nearly 500 species are found all over the world. The origin of weed is Jamaica and Mexico. These species are widely distributed in Tropical America, Europe, Africa and Asia. Most of the researchers opined that Mexico is considered to be its native place. At first instance, the weed was noticed in Southern California of U.S.A. during 1908. One decade later, the plant spread like wild fire in mountains of northern side of the Los Angeles basin. *Eupatorium adenophrum* flourishes in areas having year round rainfall. It dominates the native vegetation aggressively within a short time.

17.5.1 *Entrance in India*

Eupatorium was introduced in Indian subcontinent in 1950 and spread rapidly in the later part of the century. The weed is explosively spread in North-Eastern states. The weed is continuously invading in Nagaland since 25 years. It eclipsed the natural grasslands and meadows situated in between1000 m to 2100 m altitudes. It also colonized rapidly in newly cleared forest areas of Himachal Pradesh, Jammu & Kashmir, Uttar Pradesh and Bihar. *Eupatorium* has speedily invaded wastelands, raodsides, railway tracks, abandoned fields, sides of irrigation channels in North-eastern States. It quickly covered thin pine stands in the hilly regions of Assam. In South, the weed infested heavily in the hills of Tamil Nadu.

17.5.2 *Possible use of manure*

The weed being succulant is easily decomposed and form into good organic manure that can be used in various crops. At young stage, the loppings are used as green leaf manure in paddy fields. The leaf manure should be incorporated into the soil while pudding the field. This gives beneficial effect on rice crop performance and yield. In row crops it is used as a mulch which not only suppress the weeds but also arrest the moisture loss from the soil.

Another advantage of the weed is its inhibitory effect on

most obnoxious weed *Lantana camara*. *Eupatorium* arrest the germination of *Lantana* through allelopathic effect and reduce it infestation.

17.5.3 *Chemical means*

Spraying with glyphosate @ 2.25 to 4.50 kg/ha result in complete killing of this obnoxions weed. Aged stands can be controlled effectively with the application of one per cent glyphosate solution.

In some places, the weed is removed mechanically as its root system is not so firm in the soil. In hilly areas like Nilgiris and Darjeeling, gall fly insect *Procecidochares utilis* feeds on the weed and controls to some extent. The desired level of success can be obtained in low to moderate rainfall areas with this biological control. The weed assumes alarming proportion in high moisture zones and difficult to control with biological means.

17.6 Monster from the murky lands : Water hyacinth

The beautiful water hyacinth (*Eithhornia crassipes*) is most troublesome aquatic weed in the tropical third world. The weed is present in free-flowing or stagnant waterbodies or anchored to the hydro-soil or emergent in the drainage ditches. It is a perennial type with high reproduction rate at 8000 in one season. Water hyacinth multiplies at five per cent per day in stagment waters and double its population for every 15 to 18 days under ideal conditions. The foliage is flat on the surface of water to form some sort of blanket or mat.

Water hyacinth is able to expand very fast during the rainy season because it has prolific vegetative propagation from plants surviving in the remaining water bodies left in the dry season. It has a good competitive advantage over most of other aquatic plant in marshy environment. It completely dominates in periodically flooded landscapes with warm temperature and abundant insolation.

17.6.1 *Environmental and health hazards*

Hyacinth blocks water bodies and irrigation systems and interfere with water flows. It lesson the strength of concrete structures and bridges and cause damage greatly. It also causes water logging, salinity in surrounding lands and creates false surface on water resulting in trapped animals. The weed decrease the water quality and lesson the potable water. It also foster malaria, river blindness, encephalitis and yellow fever by harbouring mosquitos.

If left unchecked it may choke big reservoirs and canals and causes immense damage to the ecosystem. Everyday about five hectares of infested area gets added for one hundred acres of existing area infested with hyacinth.

17.6.2 *Hyacinth shifts in tropical waters*

Waters hyacinth is a native of South America and harmless plant species in the native continent. Its introduction in the old world caused important shifts in the aquatic weedflora in many of the Asian countries due to favourable environment. The plant is introduced in Bengal around 1890 and infested very rapidly throughout India. Infestation is very heavy in the plains around Ganga and Brahmaputra rivers.

The irrigation and drainage channels of every town and villages were densely infested with the weed and restricts the flow of water. This is causing huge losses of irrigation water and electricity. Even Delhi Municipal Corporation spending more than Rs. 30 lakhs every year for removing the hyacinth cover from drainage canals.

17.6.3 *Problems with chemical control of water hyacinth*

Clearing localised infestations of water hyacinth manually is an effective method of controlling the weed. Reinfestations are possible if the disposal of weed is irregular. However, using manual labour is highly labourious, expensive and uneconomical when hundreds of hectates of waterbodies to be cleared.

Immediate reinfestation is possible from the dismembered plant parts like root, stem and rhizomes.

Numerous inorganic chemical weed killers like copper sulphate, sulphuric acid, formalin, liquid ammonia, arsenic oxide and sodium chloride against water hyacinth are available. Organic herbicides like 2, 4-D, MCPA, 2, 4, 5-T, Diquat, paraquat are recommended for quick killing of the weed. Vast areas of infested water bodies can be cleared with less cost. Repeated applications will be essential to take care of reinfestations.

Use of chemicals in waterbodies may create pollution problems. Immediate and cumulative toxicity to fish and other forms of life in the aquatic habitat is possible. Water quality will be deteriorated with persistent herbicides. Dispersal of toxic chemicals through foodchains create ecological disturbances.

17.6.4 *Biological means*

Biological control by using host-specific insects and othe organisms is an effective method which offer long term potential for controlling hyacinth mats. It is relatively cheap and harmless to environment. Water hyacinth is not considered as problematic weed in its native home in South America due to its natural control through native fauna. It is regularly kept under check by naturally occurring regulatory mechanisms.

Exploration for natural insect enemies of water hyacinth was started since 60s. About 70 species of arthropods are found to be feed on hyacinth. But six species were evaluated as potential control agents which are given hereunder.

Name	Family
Neochetina eichhorniae : Warner	*Coleoptera*
N. bruchi Hustache	*Coleoptera*
Sameodes albiguttalis, Warren	*Lepidoptera*
Acigona infusella, Walker	*Lepidoptera*
Cornops aquaticum, Bruner	*Orthoptera*
Orthogalumna terebrantis, Wallwork	*Orthoptera*

Out of six species *N. eichhorniae* successfully controlled the water hyacinth in U.S.A., Sudan and Australia. In Argentina, the release of *N. bruchi* was found to be successful for killing the water hyacinth stands.

17.6.5 *Pathogenic attack*

Some of the virulent pathogens particularly fungi attack the hyacinth and damage them. Important among them arc

Acremonium zonatum
Alternaria eichhorniae
Bipolaris oryzae
B. stenospila
Rhizoctonia solani
Cercospora piaropi
C. rodmanii
Uredo eichhorniae

A wettable powder formulation of *R. solani* was found to be highly effective in several countries like U.S.A. Registration for commercial use of such type of microbial herbicides was possible in near future. Manatee (*Trichechus* sp.) a water animal which feeds voraciously on aquatic weeds was tried in some countries and found useful for controlling the hyacinth vegetation. Unfortunately these animals are declared as endangered species. Chinese grass carp (*Ctenopharyngodon idella*) was failed in controlling the weed due to it non-preference.

17.6.6 *Mites can mitigate the hyacinth cover*

Some species of mites like *Orthogaluna terebrantis* lays it eggs in lower surface of young leaves of water hyacinth. The larvae comes within one week from the egg and undergo 3 nymphal stages i.e. *proto, deuto* and *tritonymphs*. Nymphal stage will be ended within two weeks. The nymphs feed on the leaf and produce galleries in the laminae. Under field conditions two lakhs of galleries per m^2 can be produced and damage the plant heavily. the weevil *N. eichhorniae* lay more eggs on the leaf in the presence of *O. terebrantis*. Combination of weevil and mite reduce the hyacinth density within a short time when compared with one *arthropod* alone.

Successful biological control of this most serious weed will permanently open up waterbodies throughout the country for economic use without chemical pollution.

17.7 Explosive colonization of Salvinia

Salvinia molesta, an aquatic fern, of the *pteridophyta* order, originates from South America was widely distributed in Southeast Asian countries by virtue of its phenotypic plasticity. It enables it to occupy a wide range of conditions. Salvinia has been distributed throughout world and found in Asia, Africa and islands of south pacific. Till to the 1930s, the plant is confined to only botanical laboratories for conducting research but it is escaped from Colombo University, Sri Lanka and rapidly spread not only in Island but also in Indian subcontinent. Dramatic colonization of the weed takes place on large expanses of water.

In coastal regions of Andhra Pradesh and Kerala, it blocks canals, rivers, ponds and ditches and compete successfully with water hyacinth. Salvinia can survive only for a short time on mudbanks of dwindling waterbodies and cannot tolerate drying.

18

FLOODING AS WEED SUPPRESSING TOOL

Unbending at the Juvenile, is impossible to suppress at shrub —*Telugu maxim.*

Flood kills weeds by excluding air from their environment. Most of the terrestrial weeds are highly susceptible to stagnant water. Flooding is a common crop husbandry method of controlling young (1-4 leaf stage) mesophytic weeds in rice fields the world over. Rice, the major food crop, is cultivated in the countries of Asia, Africa and American continents. Annual world production is about 372 million tons gleaned from about 143 million hectares. Throughout the world flooding was practised in 45.8 million hectares (32% area) for controlling weeds emerging from paddies. About 350 species of weeds infest rice fields and among them, 18 species are found to be the most pernicious. Worldwide, weeds cause an estimated loss of 74 million tons of rice every year. In addition to yield and quality losses, there are financial losses due to cost of herbicides and mechanical operations required to control weeds. It is estimated that practice of flooding saving nearly 55.8 millions tons of rice each year through suppressing the weed competition.

18.1 Inimical environment

Flooding brings about a series of physico-chemical, electro-chemical and bio-chemical changes in soil-plant relationships, completely different from normal aerated soils. Because of the demand for oxygen in the soil and slow renewal rate through the flood water, oxygen in a warm soil containing an energy source is usually depleted within a day or so after flooding. Biological reduction processes result in the accumulation of reduced chemical ions like iron and managese and products of anaerobic reduction like methane, hydrogen, carbon dioxiode, nitrogen and organic acids like lactic acid, malic acid and acetic acid which are harmful to plant roots.

Injurious effects are caused by several metabolic imbalances ultimately resulting from insufficient oxygen. The transport of cytokinin hormones is retarded from leaves and stems to root tips. Insufficient absorption of minerals, accompanied by slower photosynthesis and carbohydrate translocation, because insufficient oxygen reduces the permeability of water to roots. The supply of ATP is limited because the electron-transport system and the Kreb's cycle cannot function without oxygen. Under hypoxic and anoxic conditions, inhibition of respiration and metabolism takes place in all parts of aerobic weeds. It also disturbs Kreb's citric acid cycle in roots. Glycolysis occurs which means lysis of sugar, break down of sugar to ethanol. No upland weed is tolerant to high ethyl alcohol concentration in the plant cells. Weeds flourished in different soilwater regimes depends on various adaptic characters (Table 18.1).

18.2 Dynamic flooding

Continuous flowing irrigation is practised in inter-mountain areas, terraces of mountain valleys and hilly regions of India, Philippines, Japan and Republic of Korea. The surplus water coming from mountain tops are used for irrigating and weed control in rice in these areas. Terraces of Himalayas in Kashmir, Dehradun and Assam valleys in India and Hokkaido and Tohoku areas of central Japan are best examples of this rice cultivation through flow flooding. Although flow flooding controls the weeds to some extent, the crop is not free from weed competition. Certain weeds like *Echinochloa colona eragrostis japonica, Marsilea quadrifolia* are encouraged as the water contains abundant oxygen. This practice require 4500-5000 mm per season.

Table 18.1 : Weeds that are flowrished in various hydroecosystems

Scientific name	*Common name*	*Family*	*Life Cycle*	*Mode of reproduction*
Well aerated soils				
Amaranthus spiosus L	Spiny amaranth	Amarantha-ceae	Annual	Small seed
Ageratum conyzoides L	Poultry lice plant	Asteraceae	Annual	Slender achene
Eclipta prostrata (L) L	Eclipta	Asteraceae	Annual	Achine
Cyperus rotundus L	Purple nutsedge	Cyperaceae	Perennial	Tubers, seeds
Cynodon dactylon (L) pers	Bahama grass	Poaceae	Perennial	Runners, seed
Dactyloctenium acgypticum (L) Beauv	Crow foot grass	Poaceae	Annual	Tiny seed
Digitaria ciliaris (Retz) Koel	Crab grass	Poaceae	Annual	Narrow Caryopsis
Eleusine indica (L) Caertn	Goose grass	poaceae	Annual	Oblong caryopsis
Ischaemum rugosum salisb	Rugose grass	Poacheae	Straggling Annual	Point tipped seed
Leersia hexandra SW	Leersia grass	Poaceae	Perennial	Root stocks/ seed
Paspalum scrobiculatum L	Paspalum grass	Poaeae	Perennial	Stem bran-ches/seed
Saturation regime				
Brachiaria mutica (Forsk) stapt	Buffalo grass	Poaceae	Stout perennial	Runners/ seed
Commelina benghalensis L	Fire leaf	Commlina-ceae	Prostrate annual	Seed/under ground and stems

Contd......

Scientific name	Common name	Family	Life Cycle	Mode of reproduction
Echinochloa colona (L) Link	Barnyard grass	Poaceae	Tufted annual	Seed
Fimbristylis littoralis Gaud	Devil sedge	Cyper-aceae	Tufted annual	Tiny seed
Shallow Flooded regime				
Alternanthera sessilis (L) R. Br.	Water amaranth	Amaranth-aceae	Prostrate annual	Seed/ Vegetative branches
Cyperus difformis L	Brahmin sedge	Cyperaceae	Tufted annual	Seed
Cyperas iria L	Umbrella sedge	Cyperaceae	Tufted annual	Triangular achene
Echinochloa glabrescens Munro ex Hood	Pyramidal barnyard grass	poaceae	Stout annual	Seed
Leptochloa chinensis (L) Nees	Sprangle top	Poaceae	Stout annual	Seed
Leptochloa filiformis L	Feather grass	Poaceae	Slender annual	Seed
Monochoria vaginalis (Burms Prege)	Monochoria	Pontederia-ceae	Broadleaf annual	Tiny seed
High pond potential				
Pistia stratiotes L	Pistia	Araceae	Free floating perennial	Off-shoots and seed
Ipomoea aquatica Forsk	Water Ipomea	Convolvula-ceae	Spreading perennial	Seed and Veg. cuttings
Sphenoclea Zeylanica Gaertn	Goose Weed	Sphenocle-aceae	Flesh, hollow annual	Tiny seed
Ludwigia octovalvis (Jacq) Raven	Lowland Clove	Onagra-ceae	Shrubby annual	Seed
Eichhoria Crassipis	Water hyacinth	Pontederi-aceae	Free floating perennial	Off-shoots/ seed

18.3 Static flooding

Continuous static flooding is usually practised in all South Asian, Latin America and African countries where adequate water supplies are available. The depth of flooding varies from 5-15 cm depth and water requirement varies with 600-1500 mm per crop season. Grassy weeds are effectively controlled in shallow water regime but broadleaf weeds proliferate as usual due to certain adaptations. Certain semiaquatic weeds such as *Ipomoea aquatica, Sphenoclea zeylamica* and *Monochoria vaginalis* pose competition to the rice crop.

18.4 Cyclical submergence or rotational flooding

Cyclical flooding is practised in areas where water supplies are limited to continuous flooding. Required amount of water for flooding was given at regular intervals. At times, the field is without standing water between two irrigations, but the soil is somewhat wet enough to avoid water stress. This rotational flooding is highly effective in Taiwan, Philippines and India. Usually the system is designed to flood thousand hectares of rice at 5-day interval. It is difficult to manage the weeds in the system due to drying of soil between two irrigations. The spectrum of weed flora and biomass production of weeds is high compared to static flooding.

18.5 Germination of weed seed under flooded condition

Certain weeds like *Monochoria vaginalis* and *Echinochloa glabrescence* can germinate in marshy soils. These seeds can also germinate under as much as 10 cm of water because it require less of free oxygen for germination than do some mesophytic weed seeds.

Echinochloa sp. seed is capable of releasing oxygen by enzymatic process that occurs during germination. These seedlings can be kept alive under water for nearly 50 days before they die. Any soil cover under water reduces oxygen supply to the seed. Germination decreases as soil cover increases. Anaerobic respiration enables these weed seeds to germinate even at a low oxygen level. Seedlings of *Leptochloa chinensis* exhibits, strong hexose utilization mechanism and a highly functional fermentation system which partially explains their ability to function in flooded soils.

18.6 Coleoptile elongation of barnyard grass

Some of the wild rice and *Echinochloa Cruss-galli* seeds will germinate under hypoxia or anoxia, but they do so by upward protrusion of the radicle. Roots hardly develop, but the coleoptile continues to grow under hypoxia faster than under anoxia. Enhanced growth of coleoptile results from accumulation of ethylene gas by young weeds and this plant harmone promotes coleoptile elongation at the expense of root growth.

18.7 Defensive mechanism of weeds in flooded soils

Some weeds like *Echinochloa glabrescens* Munro ex hook f and *Marsilea minuta* L have some physiological adaptations to survive in water logged soils. These plants have fermentation system which functions, strangely enough, to compensate for the decreased aerobic respiration, a factor responsible for the ability of aquatic weeds to germinate and grow under very low oxygen concentrations. Another defensive mechanism of aquatic weeds under conditions of oxygen deficiency is that roots are lignified which tend to exclude reduced toxic substances. Waterlogging enhances manganese toxicity. These weeds are more tolerant of high manganese content than mesophytic weeds.

18.8 Energy releasing pattern of hydrophytes

Liberation of energy in hydrophytic weed roots is carried out through anaerobic respiration process unlike aerobic metabolism in other mesophytic weeds. The phenomena, under anaerobic conditions, use more sugar, produced more CO_2 and ethanol is called *pasteur effect.* The *pasteur* effect no doubt causes decreased carbohydrate reserves in flooded soils and probably helps explain why weeds with swollen rhizomes or thick roots that store foods can survive anoxia longer.

Anaerobic metabolism may result in the production of two molecules of glucose as against the yield of 686 K cal of energy and 6 mols of CO_2 in aerobic metabolism.

Aerobic respiration in mesophytic weeds

$C_6H_{12}O_6 + 6O_2 \rightarrow 6CO_2 + 6H_2O + 38ADP + 38Pi + 38\ ATP$

1 ATP = 12 kg Cals energy

$\therefore$ 38 × 12 = 465 kg Cals

Anaerobic respiration in hydrophytic weeds :

$C_6H_{12}O_6 \rightarrow 2C_2H_5OH + 2CO_2 + 2ADP + 2\ Pi + 2ATP$

$\therefore$ 2 × 38 = 78 kg Cals.

If CO_2 evolution is equal in both the cases, glucose utilization is 200 per cent greater in hydrophytic weeds is only 24 per cent as much in fermentation as in respiration.

18.9 Peculiar feature of submersed weeds

The physiology of submersed weeds present in deep water rice fields is highly interesting and curious and differed from terrestrial weeds. The submersed plants use oxygen and carbon dioxide most economically. These gases are present in minute quantities in water and 30 time less than atmospheric air. The submersed weeds have an extensive development of a system of air chambers and air spaces in their tissues so that the entire plant body can obtain enough oxygen for respiration by internal circulation. Besides, oxygen produced during photosynthesis is stored in specially designed lacunate tissues which are distributed in all parts of weed. The submersed weeds utilise the dissolved CO_2 and bicarbonates for carrying carbon assimilation process with the help of enzyme carbonyl anhydrase.

Light in water cut-off by various components of dissolved and suspended constituents. Water molecules absorb the ultraviolet and infra-red rays of sunlight while the organic solutes filter the blue and violet part. Chlorenchymatous tissue present in mesophyll helps in absorption of diffused light present in water.

18.10 Ventilation of water weeds

Absence of oxygen around the roots of hydrophytes inhibits aerobic respiration in roots and allows potentially harmful materials to accumulate in the rhizosphere. Certain aquatic weeds like *Sphenoclea zeylanica, Monochoria, Vaginalis* and *Ludwigia octovalvis* develops internal gas spaces (lacunae) in this evnironment, serving primarily to transport oxygen to buried roots and rhizomes. Oxygen is diffused to the root from photosynthesising leaves and carbon dioxide diffused from the roots toward the leaves. During night, methane that is produced in sediments, occurred throughout plants. During day, methane was present in petioles of older leaves but not in young leaves. The lacunar system between young and older leaves is continuous through which gases move freely by mass flow.

18.11 Survival of rice in submerged soils

Rice (*Oryza sativa* L) is unique among cereals in its ability to germinate and grow successfully in water submerged soils. In swamp rice, the field is inundated throughout the lifespan of the crop. There are some unique anatomical characters of rice which enable it to survive in waterlogged soils. Roots of rice plant have spongy-cells inside and the roots are filled with air. Porosity of rice roots increases with the depth of submergence. Oxygen is diffused from leaves via the tillers and stems to roots through intercellular spaces or channels in the cortex tissues. It is not known how oxygen reaches young tissues, which are usually devoid of intercellular space. Nevertheless, the mechanism is sufficient not only to meet oxygen, requirement of respiring root cells but also to secrete oxygen or oxidised compounds into the external root micro-environment and ward off the entry of reduced substances.

As a result of the exudation of oxidised compounds, rice roots become coated with yellowish red precipitates of unknown composition. It is presumed that they are formed by iron and manganese oxides and hydroxides. Root coats are not formed around root tips.

18.12 Oxidation Power of roots—The index of tolerance

Several land crops have the capacity to oxidise their root environment (rhizosphere) when flooded. The root oxidation capacities of common crops in terms of the amount of naphylamine oxidised per gram of root during flooding (Table 18.2) indicating that rice roots are having high oxidation potential than any other crop. Further one can understand why sorghum tolerates temporary flooding better than Maize.

Table 18.2 : Relative root oxidation power of some field crops under flooded condition

Crop	*Naphthylamine oxidised in 48 hr. (mg/g dry root)*
Maize (*Zea mays*)	1
Oats (*Avena sativa*)	3
Sorghum (*Sorghum bicolour*)	4
Wheat (*Triticum aestivum*)	5
Soybean (*Glycine max*)	7
Rice (*Oryza sativa*)	15

Another physiological adaptation is that acetate which is produced in root from acetyl COA is decomposed to CO_2 via glycolic acid cycle and produces H_2O_2 in the process. H_2O_2 is decomposed by a *catalase* with the production of O_2 which becomes available to roots. Rice roots thus get oxygen not only through internal ventilation system but also chemically through enzymatic process.

18.13 Prolonged flooding for perennials

Some of the pernicious weeds like Tiger grass (*Saccharum spontaneum* L), Johnson grass (*Sorghum helepense* (L) pers), purple nut sedge (*Cyperus rotundus* L), and Thatch grass (*Imperata cylindria*) which reproduces rapidly by means of high viability seeds and rhizomes are very difficult to control by any means. Manual weeding is time consuming and requires special crowbars for soil digging and collection of rhizomes. Chemical control is not effective and repeated applications are required.

18.14 Reclamation of thatch grass infested fields

For these perennials, prolonged flooding for about 2-3 months during rainy season effectively eradicate these weeds. In India, tiger grass is infested in about 3 million hectares in cropped lands. Control measures during winter is not effective because the grass goes under dormancy during the winter. In central India where irrigation sources are plentiful, the weed is controlled with some modifications. The infested fields are ploughed deeply during the summer and plot are impounded with water up to 20 cm depth for one week. Then a tooth-pegged beam is run in wet field to take out tiger grass rhizomes loosened by the flood water. This follows cultivation of wetland rice which kills rest of the rhizomes due to anaerobiosis.

18.15 Flood fallow

In some parts of Coastal India and Central India, the fields are deep flooded for about 4 months during rainy season to control heavy infestations of problematic weeds such as Bermuda grass (*Cynodon drctylon*), Lantana (*Lantana camera*), fox tail (*Setaria glauca*), and *Physalis angulata*. After the rainy season (June to October), the water to let out to sow the winter crops like wheat and chick pea. This practice is called *Haveli* system, which is very effective in eradicating, difficult to control weeds. In Eastern coast of India, Quack grass (*Agropyron repens*) one of the top ten weeds of the world infested heavily in cropped areas of uplands. The weed propagate and colonizes new areas rapidly through rhizomes or soboles. These rhizomes grow fast in all directions within a short time and chemical control is not effective. The oldest but effective method of controlling the weed is through flooding. In summer, soil levees are made in these fields and the water is inundated during rainy season. The rain water may reaches up to one metre depth in these fields. The weed along with fleshy rhizomes completely decomposed due to anaerobic environment prevailed during 2-3 months. This is surest way of controlling this perennial weed.

Appendix 1

Noxious Weeds—in brief

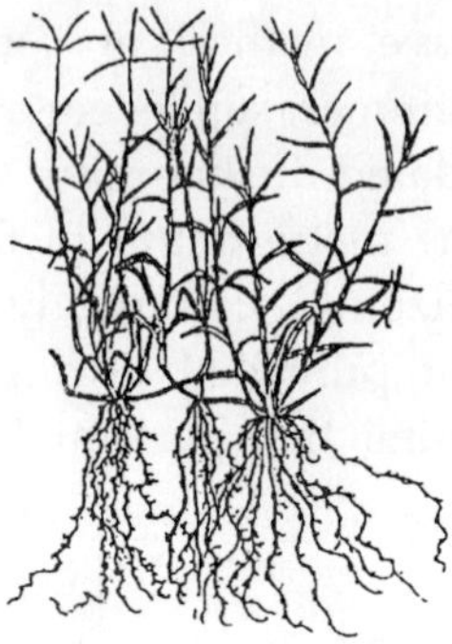

- ***Common name*** : Dog's tooth grass, Dhub grass
- ***Vernacular names*** : Sanskrit—*haritali,* Hindi—*Hariali,* Bangali—*Dubla,* Marathi—*Karala,* Tamil—*Arugum pullu,* Telugu—*Garika gaddhi,* Kannada—*Kudigarikai,* Punjabi—*talla.*
- ***Scientific name*** : *Cynodon dactylon* (L) Pers
- ***Family*** : *Poaceae*
- ***Growth habit*** : The perennial grass spreads extensively with prostrate long runners. Roots strike at nodal points. It grows to a length of 50-90 cm and forms thick mats on ground.
- ***Habitat*** : The weed is a cosmopolitan and grows well in semi-wet soils having good drainage. It has wide ecological amplitude.
- ***Propagation*** : It propagates through runners, rhizomes and seed. However, vegetative propagation is very common.
- ***Climatic diversity*** : It widely distributed throughout tropical, subtropical and semi-arid regions of the world.
- ***Crops associated*** : It is commonly found in meadows, lawns, dryland field crops, plantation crops, vegetables, and waste places.

- *Common name* : Digitaria grass, crowfoot grass
- *Vernacular names* : Hindi—*Takri,* Tamil—*Arisipillu,* Telugu—*Kakikalu gaddi*
- *Scientific name* : *Digitaria sanguinalis* scop
- *Family* : *Poaceae*
- *Growth habit* : The herb is decumbent or prostrate, short-lived perennial.
- *Habitat* : Dry soils to semi-wet soils with moderate aeration.
- *Propagation* : Through seed
- *Climatic diversity* : Warm tropical, subtropical and temperate regions.
- *Crops associated* : Sorghum, sugarcane, maize, upland rice, soybean, millets, pulses; cucurbits and vegetables.

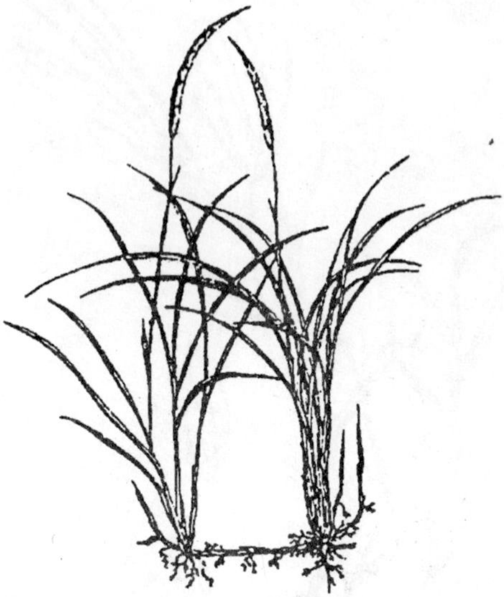

- *Common name* : Thatch grass, Cogon grass
- *Vernacular names* : Sanskrit—*Darbha*, Hindi—*Dabh*, Bengali—*Ooloo*, Telugu—*Darbha gaddi*, Tamil–*Dharbai pul*, Kannada—*Sanna dabbai hullu*, Malayalam—*Vidulam*, Punjabi—*Kusa*, U.P.—*Bharavai*, Rajasthani—*Khans*, Marathi—*Dhub*, Assam—*Batta*, Santal—*Cheru*.
- *Scientific name* : *Imuperata cylindrica* (L) Raeuschel.
- *Family* : Poaceae
- *Growth habit* : It is a perennial grass with linear and nearly upright leaves. The leaf margins have sharp edges.
- *Habitat* : It prefers poor fertile and abandoned lands where other vegetation was absent. The grass can withstand parched droughts as well as ill-drained lands.
- *Propagation* : Through rhizomes and tiny seeds travel distant places overland by speedy winds and water flows.
- *Climatic diversity* : Distributed in tropical warm regions, subtropical and temperate climates.
- *Crops associated* : Plantation crops like coffee, rubber, tea, coconut, oil palm, cincona, pineapple and all upland field crops where cultivation is minimum.

- *Common name* : Awned barnyard millet
- *Vernacular names* : Sanskrit—*Jalsamoka*, Hindi—*Samak*, Bengali—*Burashama*, Gujarati—*Abdan samo*, Marathi—*Sama*, Telugu—*Pedda wundu*, Tamil—*Oothupul*, Punjabi—*Bara sanwak*, Bombay—*Banti*.
- *Scientific name* : *Echinochloa crus-galli* (L) Beauv sp. *hispidula*
- *Family* : *Poaceae*
- *Growth habit* : Erect annual grass with stout spongy stems.
- *Habitat* : Thrives both in arable and marshy soils. Mostly found in shallow lowland paddies.
- *Propagation* : Through seed.
- *Climatic diversity* : Widely distributed in warm tropical, subtropical and cool temperate regions.
- *Crops associated* : Dryland field crops like maize, sorghum, millets and lowland rice.

- *Common name* : Little seed Canary grass.
- *Vernacular names* : Hindi—*Chiriya bajra*
- *Scientific name* : *Phalaris minor*, Retz
- *Family* : *Poaceae*
- *Growth habit* : The annual grass resembles wheat plant in size and appearance in the early stage. Stem is erect, smooth, and 0.5 – 1 metre high.
- *Habitat* : It flourishes in light soils which receives plenty of irrigation. The weed germinates in cool temperature ranging from 10 – 20°C.
- *Propagation* : It produces 300 – 500 seeds per panicle.
- *Climatic diversity* : Subtropical to temperate regions.
- *Crops associated* : Wheat, oat and barley.

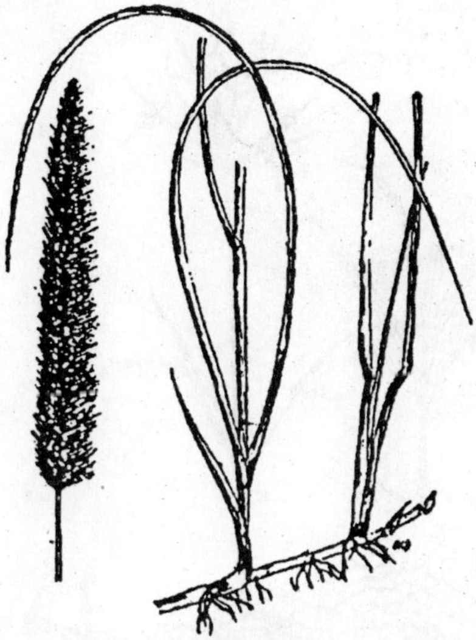

- *Common name* : Tiger grass, Kans grass
- *Vernacular names:* Sanskrit—*Kasa,* Hindi—*Kans,* Punjabi—*Kas*
- *Scientific name* : *Saccharum spontaneum,* L
- *Family* : Poaceae
- *Growth Habit* : The grass grows to a height of 4.5 metre in fertile soils. The stem is erect and leaves are linear and droopy nature. The coarse and creeping roots penetrates to a depth of 2 metres. The leaf margins are rolled up frequently.
- *Habitat* : The gaint grass mostly grown in dry and fallow soils. It grows vigorously in heavy soils where moisture retention is more. Root growth inhibits in ill drained soils.
- *Propagation* : The perennial grass rreproduces by seeds and underground rhizomes. About 90% seeds germinate under favourable conditions.
- *Climatic diversity* : Tropical and subtropical regions of the world.
- *Crops associated* : orchard crops like mango, guava, citrus, coffee, tea, sapota, amla, papaya and other waste lands.

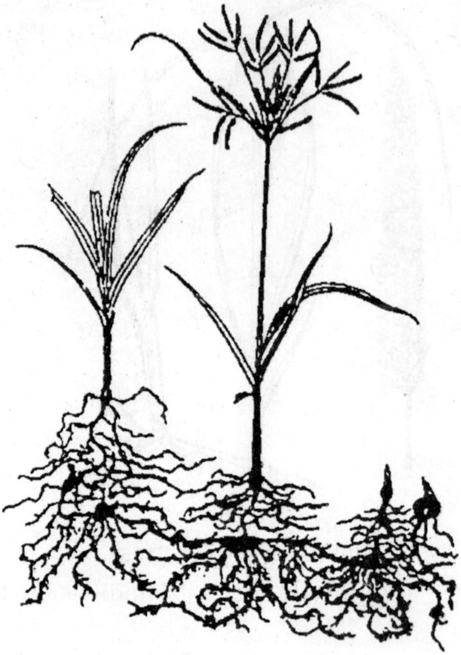

- *Common name* : Purple nutsedge, Nut grass
- *Vernacular names* : Sanskrit—*Mustaka*, Hindi—*Motha*, Bengali—*Mutha*, Marathi and Gujarati—*Motha*, Tamil—*Korai*, Telugu—*Tungamunta*, Kannada—*Tungegadde*.
- *Scientific name* : *Cyperus rotundus L.*
- *Family* : *Cyperaceae*
- *Growth habit* : It is an erect, short, perennial sedge having underground rhizomes.
- *Habitat* : Drylands with good aeration and proliferates well in semi-dry soils.
- *Propagation* : Through rhizomes, tubers and seed
- *Climatic diversity* : Tropical, subtropical and temperate regions.
- *Crops associated* : Maize, sorghum, pearl millet, finger millet, foxtail millet, sugarcane, upland rice, pigeon pea, black gram, chick pea, soyabean, cotton, brinjal and tomato.

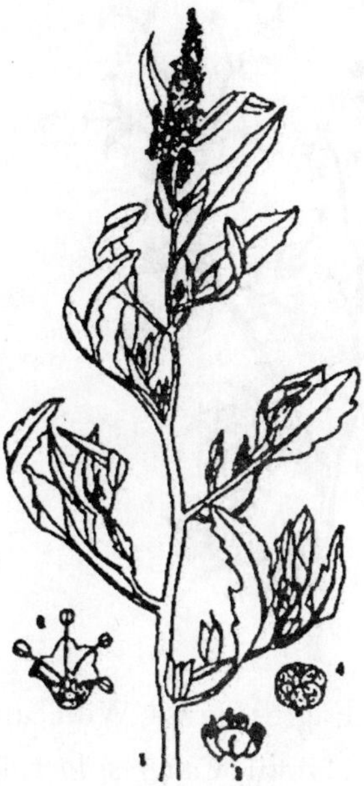

- *Common name* : Larmb's Quarter
- *Vernacular names* : Hindi—*Bethu sag*, Bengali—*Chandan bethu*, Tamil—*Parupukkirai*, Telugu—*Pappukoora*.
- *Scientific name* : *Chenopodium album* L.
- *Family* : *Chenopodiaceae*.
- *Growth Habit* : It attains to a height of one metre under favourable conditions. The crect, annual herb, branches profusely and continue its growth throughout the year.
- *Habitat* : The plant grows luxuriously in irrigated winter crops. It prefers moist and shady habitats. Luxurious growth was seen in magnesium rich soils.
- *Propagation* : Through thousands of tiny seed which have longer viability.
- *Climatic Diversity* : Subtropical regions and cool temperate zones.
- *Crops associated* : Wheat, chickpea, barley, maize, cauliflower, cabbage, potato, oats, peas and winter vegetables.

- *Common name* : Congress grass, Wild carrot weed
- *Vernacular names* : Hindi : *Congress topi*, Telugu—*Macherla Kampa*, Bengali—*Ban gazra*, Punjabi—*Congress ghas*.
- *Scientific name* : *Parthenium hysterophorus* L.
- *Family* : *Asteraceal*
- *Growth habit* : It is an erect annual, freely branched, with large number of white flowers. The roots are profusely branched. At early stages, the radial leaves are closely pressed to the ground that do not allow any plant in their midst.
- *Habitat* : The plant prefers moist, shady and nutrient rich soils. The weed is neutral both photoperiodically and thermoperiodically.
- *Propagation* : It propagates both vegetatively and through seeds (Cypsella).
- *Climatic diversity* : It has high ecological amplitude and grown in both tropics and temperate regions.
- *Crops associated* : The weed invade fastly in cotton, sugarcane, wheat, maize, sorghum, orchards, plantation crops and waste lands.

- *Common name* : The white cock's comb, Quail grass.
- *Vernacular names* : Sanskrit—*Vitunna*, Hindi—*Safed murga*, Bengali—*Swet murga*, Punjabi—*Sarwali*, Marathi—*Kurdu*, Gujarati—*Lapadi*, Telugu—*Gurugu*.
- *Scientific name* : *Celosia argentia*. L
- *Family* : *Amaranthaceae*
- *Growth habit* : It is an crect, 50-90 cm high, stout and branched stem with linear, lanceolate leaves. Inflorescence in tapering spike. Young flowers are pink.
- *Habitat* : Flourished in light textured drylands and semi-drylands, mostly in hot season.
- *Propagation* : Through compressed, subreniform seeds
- *Climatic diversity* : Tropical to sub-tropical climates
- *Crops associated* : Maize, sorghum, castor, pigeon pea, black gram, green gram, and cotton.

- *Common name* : Water hyacinth, Horse hoof
- *Vernacular names* : Hindi—*Godipana*, Bengali—*Kachuri Pana*, Telugu—*Gurrapu dekka*, Tamil—*Akasa thamarai*, Malayalam—*Kolavazha*.
- *Scientific name* : *Eichhornia crassipes* (Mart) Solms.
- *Family* : *Pontederiaceae*
- *Growth habit* : A floating plant, rooting in the mud, forming dense mats with swollen and spongy petioles.
- *Habitat* : It proliferates in drainage ditches, village ponds, sluggish streams, lakes, swamps and floating rice.
- *Propagation* : Through stolons and numerous tiny seeds
- *Climatic diversity* : Distributed well in warm tropics and subtropical regions where water is abundant.
- *Crops associated* : Floating rice, deep water rice, lowland rice and drainage ditches.

- *Common name* : Witch weed, Striga, Sorghum Jasmine.
- *Vernacular names* : Hindi, *Agia*, Marathi—*Talap*, Gujarati—*Tahuli*, Telugu—*Pogakumalle*, Tamil—*Pallipoondu*, Malayalam—*Theepalli*.
- *Scientific name* : *Striga asiatica* (L) Kuntze
- *Family* : *Scrophulariaceae*
- *Growth habit* : It is a very slender, erect, stiff, annual herb and grows to a height of 45 cm. The leaves are linear, narrow and surface is rough. It is a partial parasite on some crops.
- *Habitat* : It is a partial parasitic weed having green leaves by which they conduct photosynthesis and manufacture sugars. The seed germinates and grow towards the host root through chemotropic response. Proliferates in dry regions.
- *Propagation* : Through light, tiny seeds which are dispersed by wind.
- *Climatic diversity* : Tropics and subtropics.
- *Crops associated* : Sorghum, pearl millet, finger millet, sugarcane, rape and mustard.

Appendix 2

Cultural Weeding with Animal Operated Ploughs

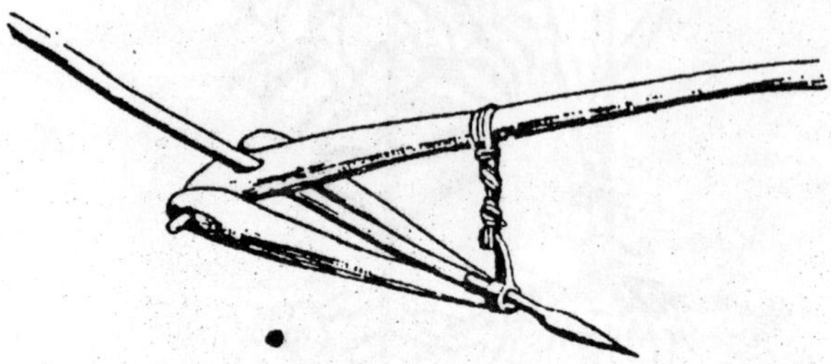

Fig. 1 : Ethiopean plough

Features

- Can be used for land preparation even in hard soil
- Helps in post-emergence weed control
- Row cultivation is possible

Country of origin : Ethiopia

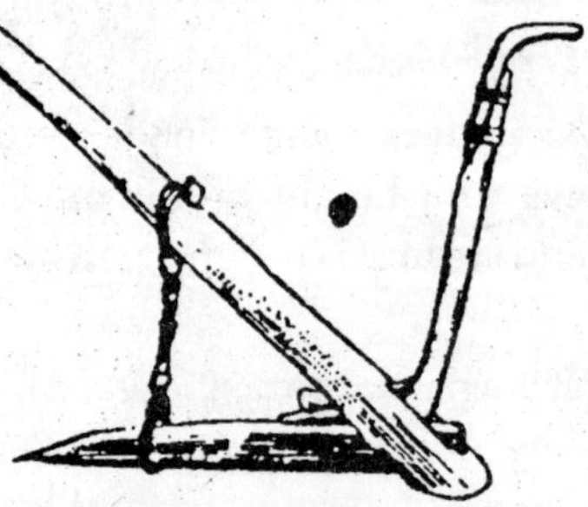

Fig. 2 : Chilean plough

Features

- Can be drawn by animals
- Row spacing should be wide enough for passing bullock
- Weeds can be uprooted in perennial weed infested fields.
- Symmetric type of breaking

Country of origin : Chili

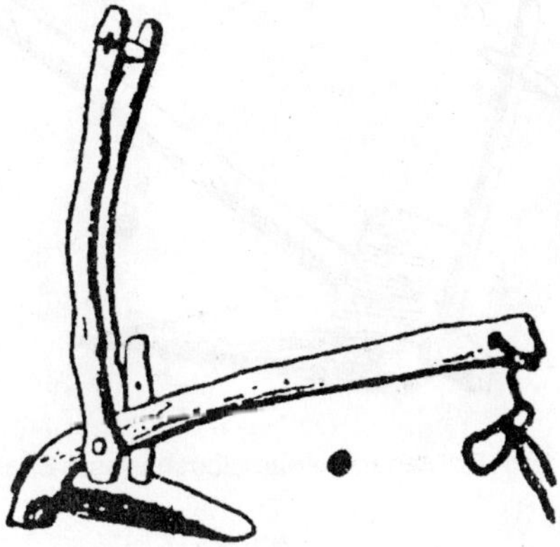

Fig. 3 : Desert plough of Basra

Features

- Disturb the soil at shallow depth only
- Highly suitable for desert sandy soils
- Locally designed for desert conditions.
- Helps in soil and water conservation

Country of origin : Iraq

Fig. 4 : Balady plough of Egypt

Features

- Symmetric breaking of soil.
- Break the soil layer but not invert the soil.
- Throws soil to both sides.
- Its beam, body and handle are made with wood.

Country of origin : Egypt

Fig. 5 : Maharashtrian plough West Coast

Features

- It is a welcome implepment for farmer with small resources.
- Suitable for operations in narrow spaced row crops.
- Can be used for sowing of row crops.

Country of origin : Peninsular India

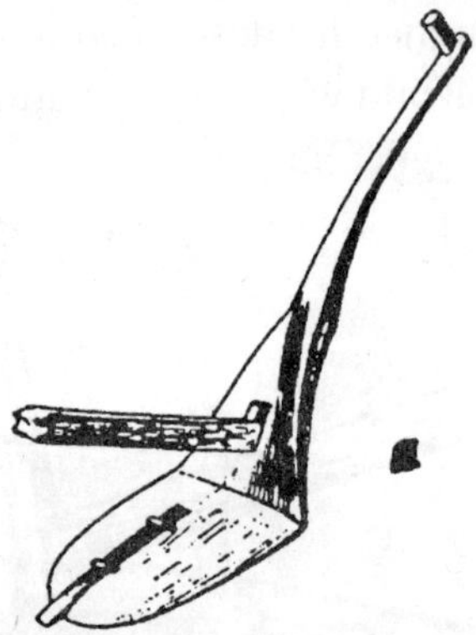

Fig. 6 : Paddy plough of Coastal Orissa

Features

- It is a bullock drawn wooden country plough.
- Used for puddling the paddy fields.
- High soil moisture or stagnant water in the field is required.
- Large area can be covered as it puddle the soil at shallow depths.

Country of origin : East Coast India

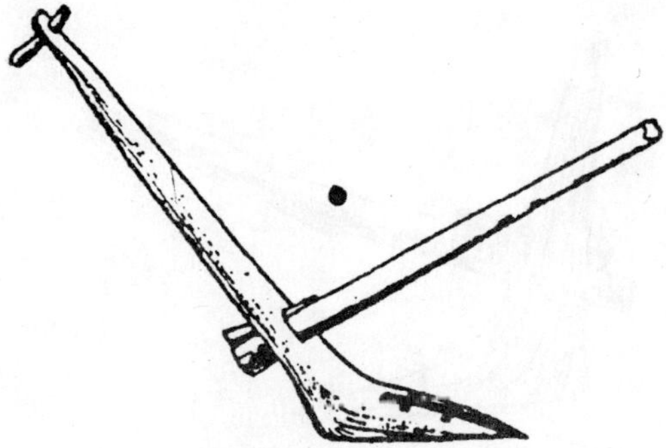

Fig. 7 : Indian desi plough of Coastal Andhra.

Features

- It is a light implement with simple design.
- Its beam, body and handle are made with wood except share point.
- Does not invert the soil.
- Burying of weeds is not possible.
- It is suitable for light soils.

Country of origin : South India

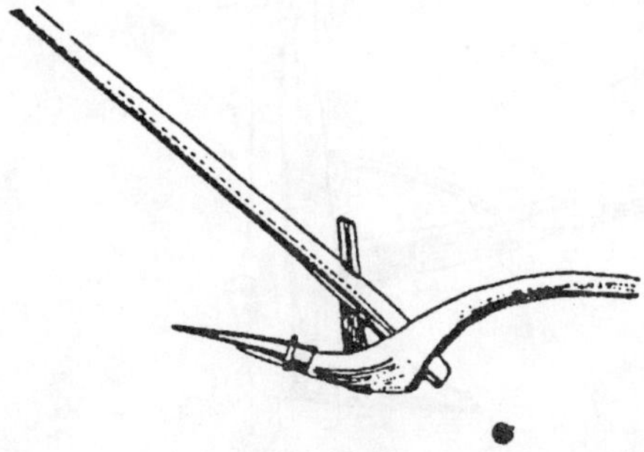

Fig. 8 : Turkey country plough

Features

- The ard produces a slightly ridged tilth.
- It has a breaking and cutting action of soil.
- Needs repeated cross ploughings to obtain good tilth.

Country of origin : Turkey

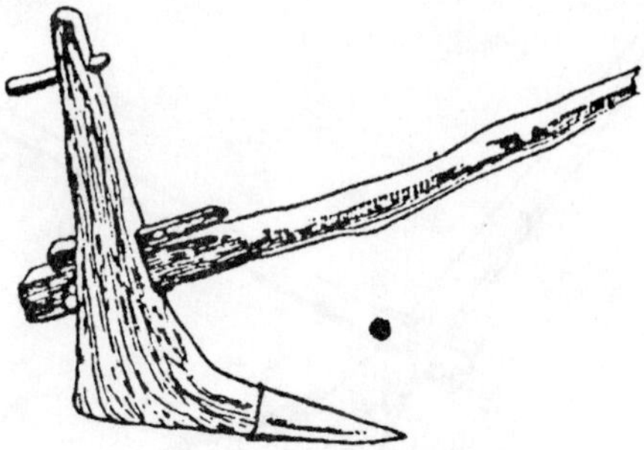

Fig. 9 : Afghanistan plough

Features

- All parts are made with wood except the iron share
- Versatile and light implement in dry zones used by nomads.
- Used for shallow tillage.
- Leaves weeds mostly uncut
- Design is simple with low cost.

Country of origin : Afghanistan and Beluchistan

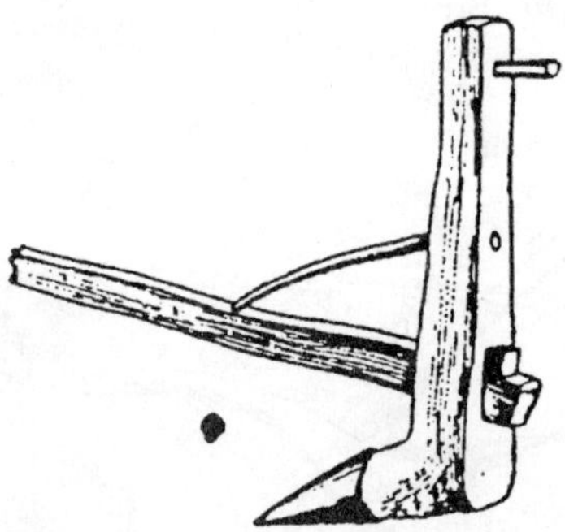

Fig. 10 : Syrian plough with metallic share

Features

- Suitable for hard and semi-arid regions
- Low working speed
- Depth adjustment is not possible
- Produce slightly ridged tilth

Country of origin : Syria

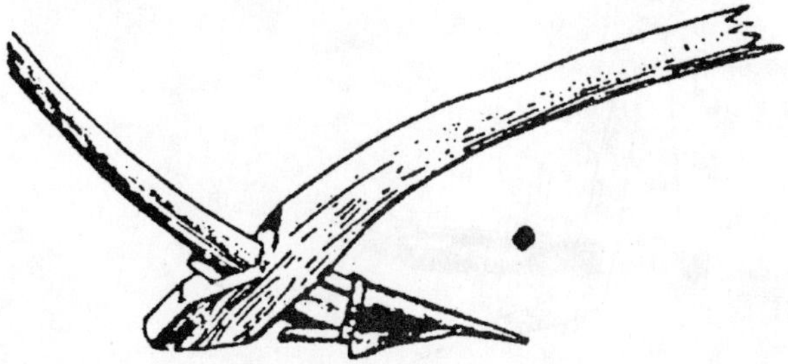

Fig .11 : Iraqi (Central) plough

Features

- Traditional animal drawn plough for land tilling
- Can be used for interculture in row crops
- Used for seed bed preparation in cereals
- Improves weeding efficiency than manual weeding

Country of origin : Central Iraq

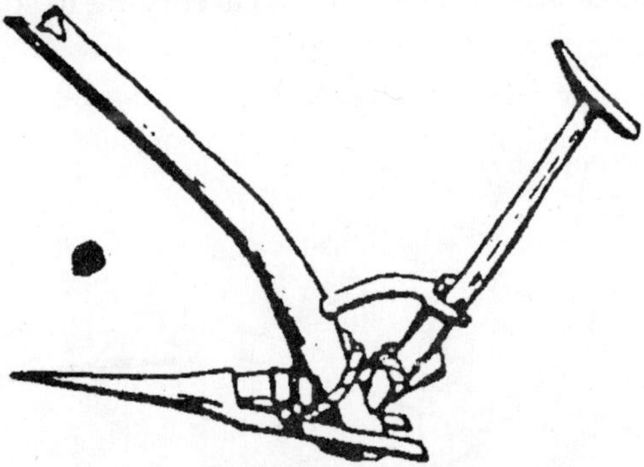

Fig. 12 : Syrian plough

Features

- Used in dryland crops to stir the soil
- Bury the young weeds by cutting their root system.
- Less strain for animals
- Designed by peasant cultivators based on long experience

Country of origin : Syria.

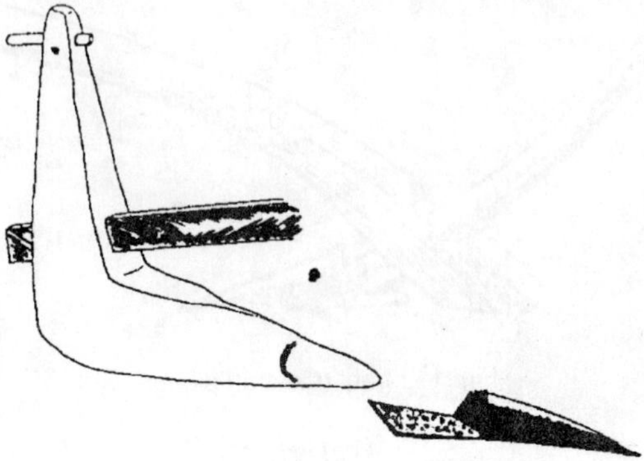

Fig.13 : Socket share plough

Features

- The plough has a socket share.
- The body ard is mostly wide spread in India
- It is stronger than other types of desi ploughs
- Socket shares can be changed according to soil types.
- Better cutting of weeds can be done.

Country of origin : India

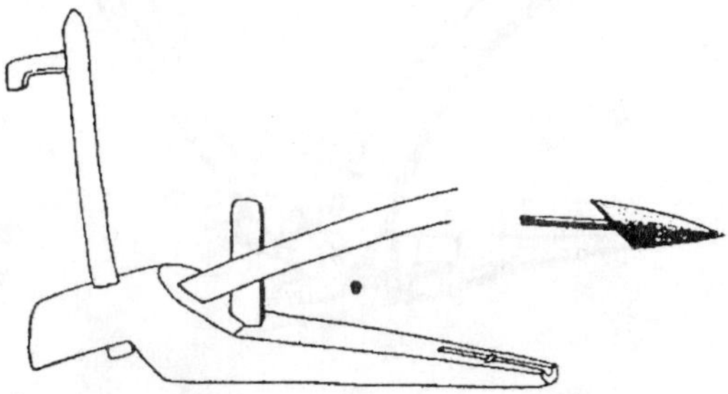

Fig.14 : Chisel share plough

Features

- The share is chisel shaped
- Mostly used in Northern India.
- Strong weeds can be culted with metal share.
- Shares can be changed depends upon the soil and weed flora.

Country of origin : India

Fig. 15 : Kurdistan plough

Features

- Easily manageable, light implement of dry areas.
- Suitable for hard soils of semi-arid regions
- More ploughings are necessary to obtain good tilth

Country of origin : Kurdistan and Iran

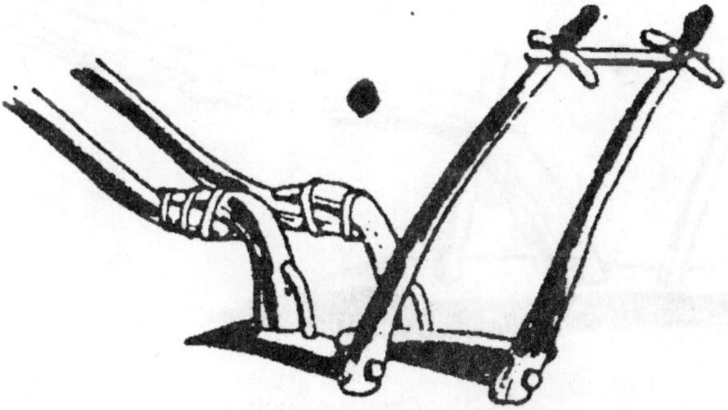

Fig. 16 : Double sole plough of Syria

Features

- Light implement of dry land cultivation.
- Working time is reduced because of having double sole.
- Low working speed.
- Inter-row weeding can be done.

Country of origin : Syria

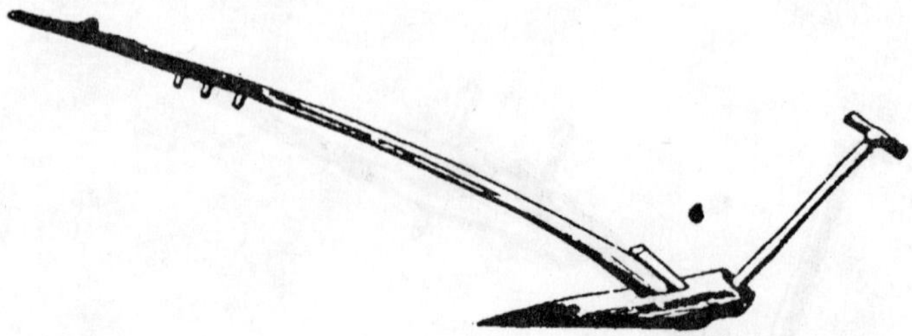

Fig . 17 : Nepali plough

Features

- Simplicity of design and ease of operation.
- Used for shallow tillage
- Weeds are uncut and unburied.

Country of origin : Nepal

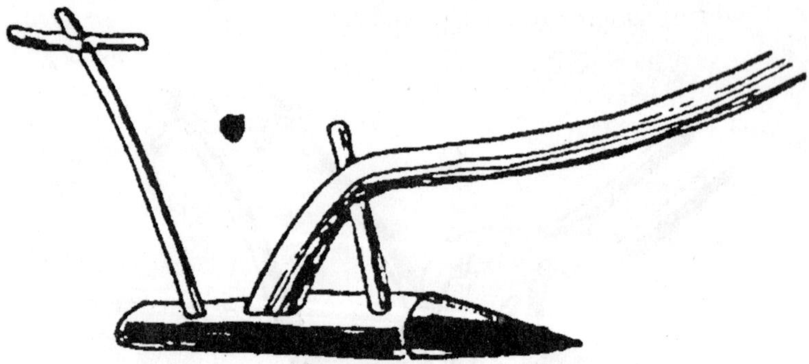

Fig. 18 : Cyprus plough

Features

- It can be used for seed-bed preparation and inter-row weeding.
- Low working speed
- Insufficient capacity to eradicate weeds
- Symmetric type of breaking

Country of origin : Cyprus

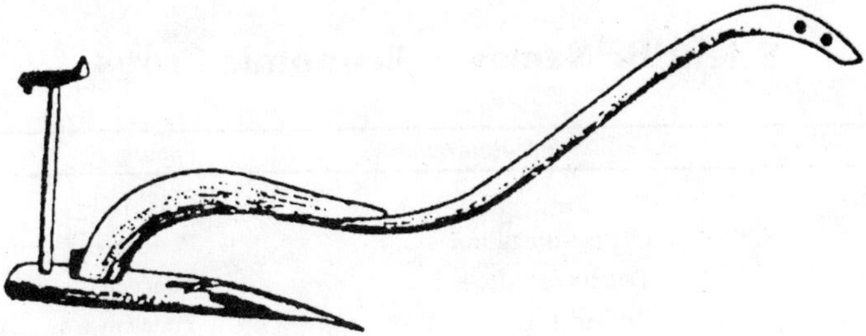

Fig. 19 : Pakistan plough

Features

- It is a versatile and easily manageable plough.
- Used by nomads for shallow tillage
- It leaves weeds mostly uncut.

Country of origin : Pakistan

Fig . 20 : Kabul plough

Features

- Light implement for dryland agriculture.
- Simplicity in design and low cost
- Suitable for hard soils of semi-arid regions

Country of origin : Afghanistan

Appendix 3

Scientific Names of Economic Crops

Name of Crop	*Scientific name*	*Family*
Cereals		
Rice	*Oryza sativa* Linn	Poaceae
Wheat	*Triticum aestivum* L	Poaceae
Maize (corn)	*Zea mays.* L	Poaceae
Barley	*Hordeum vulgare* L	Poaceae
Oats	*Avena byzantina* C. Kock	Poaceae
Rye	*Secale cereales* L.	Poaceae
Buckwheat	*Fagopyrum esculentum*	Polygonaceae
Millets		
Sorghum	*Sorghum bicolor* (L) Moench	Poaceae
pearl or bulrush millet	*Pennisetum americanum* (L) K. Schum	Poaceae
Italian or Foxtail millet	*Setaria italica* Beauv	Poaceae
Proso/Common millet	*Panicum miliaceum* Linn	Poaceae
Little millet	*Panicum miliare* Lam	Poaceae
Ragi or Finger millet	*Eleusiye Coracana* (L) *Gaertn*	Poaceae
Kodo/Ditch millet	*Paspalum scrobiculatium* L	Poaceae
Japanese barrenyard millet	*Echinochloa frumentacea* (Roxb). Link	Poaceae
Shama millet	*Echinochloa colona* (L) Link	Poaceae
Australian millet	*E.chinochloa decompositum* L.	Poaceae
Browntop millet	*Brachiaria ramoa* (L) Stap	Poaceae
Teff millit	*Eragrotis tef (zucc)* Trotter	Poaceae
Hungry rice	*Digitaria iburua* Stapf	Poaceae
Fonio rice	*Digitaria exilis* sfaph	Poaceae
Job's tears	*Coix lacchryma* jobi Linn	Poaceae
Wild rice	*Zizamia aquatica* L	Poaceae
Pulses		
Bengal gram	*Cicer arietinum* L	Fabaceae
Pigon Pea	*Cajanus cajan* Milsp	Fabaceae

Name of Crop	*Scientific name*	*Family*
Black gram (urd)	*vigna mungo* (L) Hepper	Fabaceae
Green gram (*Moong*)	*Vigna radiata* (L) Wilczek	Fabaceae
Moth bean	*Vigna aconitifolia* (Jacq) Marechal	Fabaceae
Lentil (Masur)	*Lens culinaris* Medic	Fabaceae
Peas (matar)	*Pisum arvense* L.	Fabaceae
Chick ling vetch (Khe sari)	*Lathyrus sativus* L	Fabaceae
Cowpea	*Vigna unguiculate* (L) Walp	Fabaceae
Cluster bean (guar)	*Cyamopsis tetragonoloba* Taub	Fabaceae
French bean	*Phaseolus vulgaris* L	Fabaceae
Indian bean	*Lablab purpurius* (L) Sweet	Fabaceae
Soybean (bhat)	*Glycine max* L	Fabaceae
Broad bean	*Vicia faba* L	Fabaceae
Velvet bean	*Stizolobium deeringianum* Brot	Fabaceae
Rice bean	*Vigna umbellata* (Thumb) Ohwi and Ohashi	Fabaceae
Cat greengram	*Vigna trilobata* (L) Verd Court	Fabaceae
Limabean	*Phaseolus lunatus* L	Fabaceae
Common vetch	*vicia sativa* L.	Fabaceae
Vetches ***for pastures and green manures***		
Hairy vetch	*Vicia chilosa* L.	Fabaceae
Smooth vetch	*Vicia villosa* L	Fabaceae
Purple vetch	*Vicia atropurpurea* L	Fabaceae
Narrowleaf vetch	*Vicia angustifolia* L.	Fabaceae
Woolypod vitch	*Vicia dasycarpa* L	Fabaceae
Bitter vetch	*V. eruilla*	Fabaceae
Monantha vetch	*V. monantha* L	Fabaceae
Hungarian vetch	*V. pannonica* L	Fabaceae
Bard vetch	*V. Calcarata* L	Fabaceae
Augta vetch	*Phascolus angustufolia* L	Fabaceae
Big flower vetch	*Phaseolus grandiflora* L	Fabaceae
Crown vetch	*Corronilla varia* L	Fabaceae
Cyperus vetch	*Lathyrus ochrus* L	Fabaceae
Genge	*Astragalus cicer* L	Fabaceae
Joint Vetch	*Aeschynomene americana* L	Fabaceae
Kidney vetch	*Anthyllus vulneraria* L	Fabaceae

Name of Crop	Scientific name	Family
One-leaved vetch	*Vicia articulata* L	Fabaceae
Vetchling	*Lathyrus cicera* L	Fabaceae
Garden bean	*Vicia faba* L	Fabaceae
Oil Seeds		
Sesamum	*Seasamum indicum* L	Pedaliacease
Groundnut	*Arachis hypogaea* L	Fabaceae
Sunflower	*Helianthus annus* L	Asteraceae
Soybean	*Glycine max* Merril	Fabaceae
Safflower	*Carthanus tinctorius* L	Asteraceae
Niger	*Guizotia oleifera* DC	Asteraceae
Coconut	*Cocos nucifera* L	Arecaceae
Castor	*Ricinus communis* L. F. Br	Euphorbiaceae
Rape (Yellow sarson)	*Brassica campestris* L.	Brassicaceae
Brown sarson	*Brassica compestris* L. Vari brown	Brassicaceae
Toria	*Brassica campestris* L. Var. Toria	Brassicaceae
Mustard	*Brassica juncea* Czerm and Coss	Brassicaceae
Banarasi Rai	*Brassica nigra* (L) Koch	Brassicaceae
Taramira	*Eruca sativa* L	Brassicaceae
Linseed	*Linum usitatissimum* L	Limaceae
Oil Palm	*Elaesis guineensis* Jaco	Arecaceae
Olive	*Olea europaea* L	Oleaceae
Cacao Butter	*Theobroma cocoa* L	Sterculiaceae
Shea butter	*Butryospermum paradoxum* (Gaertn Hepper)	Sapotrceae
Essential Oils		
Rose oil	*Rosa damascena*, Mill	Rosaceae
	Rosa alba L.	Rosaceae
	Rosa centifolia L	Rosaceae
Geranium	*Pelargonium capitatum* (L) Ait	Geraniaceae
	P. radula L	Geraniaceae
	P. graveolens L' Herit	Geraniaceae
	P. odoratissium (L) Ait	Geraniaceae
	P. roseum L.	Geraniaceae
Lemon grass	*Cymbopogon flexuosus*, Stapf	Poaceae
	C. citratus Stapf	Poaceae

Name of Crop	*Scientific name*	*Family*
Citronella	*Cymbopogon nardus* (L) Rendle	Poaceae
Macassar	*Schleichera trijuga* Willd	Sapindaceae
Palmarosa	*Cymbopogon martini* (Roxb) Watse	Poaceae
Vetiver	*Vetiveria Zizaniopides* (L) Nosh	Poaceae
Sandal wood	*Santalum album* L	Santalaceae
Orange blossom	*Citrus aurantium* L.	Rutaceae
Acacia	*Robinia pseudacacia* L.	Fabaceae
Ambrette	*Hibiscus abelmoschus* L.	Malvaceae
Basil	*Ocimum basilicum* L.	Lamiaceae
Cyclamen	*Cyclamen europaecum* L.	Primulaceae
Lavander	*Lavendula officianlis*, Chaix	Lamiaceae
Linoloe	*Ocotea rodioci (*Schomb*)* Mez	Lauraceae
Patchouly	*Pogostemon perilloides* (L) Mansf	Lamiaceae
Rosemary	*Rosemarrium officinalis* L.	Lamiaceae
Saffron	*Crocus sativus* L.	Oridaceae
Sugar Crops		
Sugar cane	*Saccharum officinarum* L.	Poaceae
Sugarbeet	*Beta vulgaris* L.	Chenopodiaceae
Sugar Plam	*Arenga pinnata* (Wurmb) Merril	Arecaceae
Palmyra	*Borossur flabelifer* L.	Arecaceae
Sago Palum	*Metroxylon sago*	Aricaceae
Tuber Crops		
Arrow head	*Arracacia xanthorrhiza*, Baner	Apiaceae
Arrow root	*Maranta arundinacea*, L	Marantaceae
Cassava	*Manihot esculenta* Crantz	Euphorbiaceae
Chinese water Chestnut	*Elecocharis dulcis* (Burn f.) Trin	Cyperaceae
Elephant yam	*Amorphophallus campanulatus* (Roxb) Blume	Araceae
Giant Taro	*Alocacia macrorrhiza* (L) Sccott	Araceae
Jerusalem Artichoke	*Helianthus tube* rosus L.	Asteraceae
Kudzu	*Pucraria lobata* (Willd/Ohir)	Fabacear
Lotus	*Nelumbo nucifera* Gaertn	Nymphaeaceae
Oca	*Oxalis tuberosa* Molina	Oxalidaceae
Potato	*Solanum tuberosum* L.	Solanaceae

Name of Crop	*Scientific name*	*Family*
Swamp Taro	*Cyrtosperma chamissonis* (Schott)	Araceae
Sweet polato	*Ipomoca batatas* (L) Lamm	Convolvulaceae
Taro	*Colocasia esculenta* (L) Schott	Araceae
Beet root	*Beta vulgaris* L	Chenopodiaceae
Greater yam	*Dioscorea alata* L	Discoreaceae
Intoxic yam	*D. hisphida* L	Discoreaceae
Lesser yam	*D. esculenta* (Lour) Burn	Dioscoeaceae
Potato yam	*D.* L.	Dioscoeaceae
White yam	*D. rotundata* bulbifera L	Dioscoeaceae
Yellow yam	*D. cayenesis* Lam	Dioscoeaceae
Fibre Crops		
Cotton	*Gossipium arboreum* L	Malvaceae
Flax	*Linum usitatissiumum* L.	Linaceae
Jute	*Corchorus capsularis* L.	Tiliaceae
Kenaf	*Hibiscus cannabinus* L	Malvaeae
Sunnhemp	*Crotalaria juncea* L.	Fabaceae
Ramie	*Boehmeria nivea* Gaud	Urticaceae
Mamila hemp	*Musa textilis* Nees	Musaceae
Hemp	*Cannabis sativa* L	Cannabinaceae
Sisal hemp	*Agave sisalina* perr	Amaryllidceae.
Roselle	*Hibiscus sabdariffa* L	Malvaceae
Congo jute	*Urena lobata* L	Malvaceae
Species		
Black pepper	*Piper nigrum* L.	Piperaceae
Saffron	*Crocus sativus* L	Iridaceae
Clove	*Eugenia caryophyllus* (Spregal) Bull	Myretaceae
Large Cardamom	*Aframomum meligueta* (Roscoe)	
Small Cardamum	*Elettaria cardamom* Maton	
Ginzer	*Zinziber officinale* L.	Zingiberaceae
Turmeric	*Curcuma longa* L	Zingiberaceae
Cinnamon	*Cinnamomum zeylanicum* Blune	Lauraceae
Tejpat	*Cinnamomum tamala* Ness and E berm	Lauraceae
Fennel	*Foeniculum vulgare* Mill	Apiaceae
Nutmeg/Mace	*Myristica fragran* Houtt	Myristicacae

Name of Crop	*Scientific name*	*Family*
Fenugreek	*Trigonella foenum-graccum* L	Fabaceae
Anise	*Pimpinella anisum* L	Apiaceae
Cumin	*Cuminum cyminum* L	Apiaceae
Celery	*Apium graveolens* L	Apiaceae
Dill	*Anethum graveolens* L	Apiaceae
Asafoetida	*Ferula asafoctida* Regal	Apiaceae
Black jeera	*Nigella sativa* L	Ranunculaceae
Spear mint	*Mentha spicata* L	Lamiaceae
laurel	*Laurus nobilis*	Lamiaceae
Curry leaf	*Murraya koenigii* Spreng	Rutaceae
Horse mint	*Mentha longifloia*, Huds	Lamiaceae
Basil	*Ocimum basilicum* L	Lamiaceae
Somalata	*Ruta graveloens* L	Rutaceae
Beverages		
Coffee	*Coffea arabica* L	Rubiaceae
Tea	*Camellia sinensis* (L) O. Kuntze	Theaceal
Cocoa	*Theobroma cacoa* L	Sterculiaceae
Mate	*Ilex paraguariensis* St. Hill	Aquifoliaceae
Kava	*Piper methysticum* L.	Piperaceae
Hops	*Humulus lupulus* L	Cannabinaceae
Dyes and tanins		
Madder	*Rubia cordifolia* L	Rubiaceae
Nagar motha	*Mariscus cyperinus*L	Cyperaceae
Catechu	*Acacia catechu* wild	Mimosaceal
Arusa	*Justicia adhatoda* L.	Acanthaceae
Rose lichen	*Parmelia kamst chadalis*, Ach	Parmeliaceae
Indigofera	*Indigofera tinctoria* L.	Fabaceae
Egyptian indigo	*I. articulata* Gouan	Fabaceae
Wiry indigo	*I. aspalathoides* vaul ex. DC	Fabaceae
Trailing indigo	*I. endecaphylla* Jacq	Fabaceae
Himalyan indigo	*I. gerardiana* Wall. e. Baker	Fabaceae
Hairly indigo	*I. Hirsuta* L	Fabaceae
West Indian Indigo	*I. suffruticosa* Mill	Fabaceae
Natal indigo	*I. arrecta* Hochst.	Fabaceae
Sumatra indigo	*I. sumatrana* Gaertn	Fabaceae
Vasuka indigo	*I. enneaphylla* L	Fabaceae

Appendix 4

Problematic Weed Seed Present in Crop Seeds

Weed species	*Mixed in crop seed*	*Areas of heavy infestation*
Chicorium intybus (Chicory)	Berseem	Gwalior, Indore, Bhopal, Kota and Chittorgarh
Cuscuta spp (Golden Vine)	Lucern	Ajmer and Himmatnagar
Vicia sativa	Lentil, pigeon pea	Kanpur, Shahjahanpur, Meerut, Gorakhpur, Niwari Rudrapur, Bhopal, Indore
Melilotus alba	Fenugreek (*Methi*)	Sriganganagar, Kota, Himmatnagar
Phalaris minor	Wheat	Meerat, Delhi, Hisar, Ludhiana
Echinochloa colona	Rice	Godavari, Krishna, Guntur, Nellore, Coimbatore DTs.

Appendix 5

Some Exotic Weeds Introduced into India from Various Countries

Weed Species	*Original place*
Acanthospermum australe	South America
Acrachne racemora	South America
Adinosterma lavenia	South America
Alhagi camelorum	South America
Alternanthera achyrantha	Tropical America
A. pungens	Tropical America
A. repens	South America
Aloysia triphylla	Brazil
Amaranthus spinosus	Brazil
Ammania baccifera	Brazil
Ammi huntii	Europe
Anagallis arvensis	Europe
Anotis numularia	Europe
Apium tenvifolium	Europe
Argemone ochrolenca	Mexico
Caesulia axillaris	Mexico
Calceolaria maxicana	Mexico
Carthamus oxycantha	South America
Casia occidentalis	South America
C. soppera	South America
C. tora	South America
Chenopodium amrosiodes	Mexico
Chloris inflata	Tropical Africa
Coronopus didyma	Tropical America
Croton sparsiflorus	West Indies
Datura metel	Tropical America
Desmanths virgatus	Tropical America
Eupatorium cannabinum	West Africa
E. odoralum	Jamaica
Flaveria australasica	Australia
Jatropha gossypifotia	Brazil
Lantana mista	Central America
Lippia javanica	Tropical America
Nicandra physaloides	Peru
Ocimum americanum	North America
Opuntia dillenii	South America

Appendix 6
Weeds Identified as Quarantine Pests

Abutilon theophrasti, Medic
Allium vineale. L
Ammi visnaga (L) Lam
Astragalus cicer. L
Avena sterlis. L
Brachiaria platiphylla, Griseb
Brassica kaber, DC
Bromus diandrus, Roth
Bromus rigidus, Roth
Breomus recalinus. L
Centrwrea diffusre. L
Chondrilla juncea.
Cynoglossum officinale
Cyperus amuricus Maim
Desmodium tortuosum L.
Emex australis. Steinh
Emex spinosa (L) Campdera
Froclichia floridana L.
Helianthus petiolaris. Nutt
Hyoscymus niger L.
Hypericum perforatum L.
Ipomoea coccinea L.
Laxmiumm amplexicanle L.
Leersia japonica. Honda and Honda
Lolium rigidum Gandich
Matricaria perforatum L
Mimosa pigra L.
Mollugo verticillata. Roxb.
Nassella trichotoma (Nees) Hack
Orobanche cumana. Wallr
Panicum taxanum. Buckl
Phalaris paradoxa. L
Polygonum covolvulus L.
Polygronum lapathifolium L.
Rapanus raphanistrum L.
Rumex crispus. L
Salsola vermiculata. L
Senecio vulgaris. L.
Striga hermonthica (Del) Benth.

Appendix 7
Prominent Weeds of India

Cool Weather weeds (Winter season)

Chenopodium album L
Convoluulus arvensis L
Cyperus rotundus L.
Amaranthus viridis L.
Portulaca oleracea L
Coronopus didymus (L) Sm
Euphorbia thymaefolia L
Rumex dentatus L
Melilotus alba Des
Phyllanthus niruri L.
Portulaca qyuadrifida L.
Avena fatua L.
Trigonella polycerata L.
Fumaria parviflora. Lank
Lipidium sativum. L.
Anagalis arvensis L.
Gallin sogaparviflora L
Polygonum aletum L.
Spergula arvensis L.
phalaris minor L.
Avena ludoviciana. Dur
Lathyrus aphaca L.
Vicia sativa L.
Melilitus indica L.
Ageratum conyzoides L.

Hot Weather or rainy season weeds

Cynodon dactylon L.
Digitaria sanguinalis (L) Scoop.
Setaria glauca (L) Beauv
Tribulus terrestris L
Eclipta alba L
Cannabis sativa L
Portulaca oleracea L
Euphorbia hirta L

Echinorchloa crus-galli (L) Beauv
Sorghum helepense (L) Pers
Panicum dichotomiflorum Mixchx
Cyperus esculentus L
Cirsimum arvense L.
Launea pinnatifida L
Trianthema monogyna
Imperatr cylindrica
Digera arvensis L
Ageratum conyzoides L
Phyllanthus niruri L
Ischaemum rugosum. Salisb
Fimbristylis littoralis. Gand.
Alternanthera sessiis (L) Roth
Amaranthus retroflexus L.
Convolvulus arvensis L
Celosia argentia L.
Commelina benghalensis L.
Corchorus actuangulus. Lank
Solanum nigrum L.
Eleusine indica (L) Gaertn.
Gynandropsis gynandra briof.
Physalis minima L
Crotolaria striata DC (poir)
Setaria pumila R and S
Ludwigia octovalvis (Jacof) Raven sub-sp sessiflora.

Appendix 8

Ability of Tolerance of Wild Rices to Biotic and Abiotic Stresses

Oryza species	*Resistance to pests/diseases*
Oryza nivara	Grassy stunt virus
Oryza longistaminata	Bacterial leaf blight
Oryza officinalis	Bacterial leaf blight
Oryza minuta	Bacterial leaf blight
Oryza latifolia	Bacterial leaf blight
Oryza australiensis	Bacterial leaf blight
Oryza brachyantha	Bacterial leaf blight
Oryza minuta	Blast disease
Oryza officinalis	Brown plant hopper (B P H)
Oryza minuta	Brown plant hopper (B P H)
Oryza latifolia	Brown plant hopper (B P H)
Oryza australiensis	Brown plant hopper (B P H)
Oryza granulata	Brown plant hopper (B P H)
Oryza officinalis	White backed plant hopper (W B P H)
Oryza brachyantha	Yellow stem bores
Oryza ridleyi	Yellow stem borer
Oryza minuta	Sheath blight
Oryza rufipogon	Tungro virus
Oryza officinalis	Tungro virus
Oryza rufipogon	Acid sulphate soils
Oryza rufipogon	Tolerance to waterlogging and elongation ability under blooded conditions.

Appendix 9

Conversion Factors and Metric Measures

Length :

1 Centimetre (cm) = 10 millimetre (mm)
1 Decimetre (dm) = 10 cm
1 metre (m) = 10 dm
1 decametre (dam) = 10 m
1 hectometre (hm) = 10 dam
1 kilometre (km) = 10 hm,

Conversion factors

1 inch = 2.5400 cm
1 foot = 30.480 cm
1 yard = 91.4399 cm
1 metre = 39.370 in
1 metre = 1.094 yards
1 kilometre = 0.621 miles.

Weight

1 centigram (cg) = 10 milligrams (mg)
1 decigram (dg) = 10 cg
1 gram (g) = 10 dg
1 decagram = 10 g
1 hectogram (hg) = 10 dag
1 kilogram (kg) = 10 hg

Conversion factors

1 gram = 0.03527 ounces
1 oz = 28.4
1 kg = 2.2 pound (lb)
1 lb = 454 g
1 metric ton = 1000 kg = 0.984 ton

Capacity

1 centilitre (cl) = 10 milliliters (ml)
1 Decilitre (dl) = 10 cl
1 litre (l) = 10 dl

1 decalitre (dal)	= 10 l
1 hectolitre (hl)	= 10 dal
1 kilolitre (kl)	= 10 hl

Conversion factors

1 point	= 0.561 litre
1 Quart	= 1.136 l
1 gallon (imperial)	= 4.546 l
1 fluid ounce	= 23.38 cc
1 litre	= 0.22 gallons = 1.76 pints
1 ounce	= 28.4 ml
1 imperial gallon	= 1. 2 U.S. gallon

Temperature

Centigrade °	= (F° – 32) × 5/9
Fahrenheit	= (C° × 9/5) + 32

Conversion ratio between raw to processed products

Rice	= 77% of paddy
Mustard oil	= 33% of seed crushed
Cake to seeds crushed	= 67%
Linseed oil to seed	= 27%
Cotton lint	= 1/3 of **kapas**
Cotton seed	= 2/3 of **kapas**
One bale of cotton	= 180 kg
Groundnut kernel to nuts	= 70%
Groundnut oil to nut	= 28%
Groundnut oil to kernels	= 40%
Groundnut cake to kernel	= 60%
Sesamam oil to seed	= 40%
Jaggery from sugarcane	= 10%
Crystal sugarcane Gur refinery	= 62.4%
Khandasari sugar from Gur	= 37.5%
Crystal sugar from cane	= 9.97%
Molasses from cane	= 3.5 %
Cane trash from cane harvest	= 10%
One bale of jute	= 180 kg
100 yards of hessian	= 2 times of cotton lint

Appendix 10

Ready reckoner for preparing upto 10 litres of herbicide solution of required strength

Rate of herbicide required (kg a.i. tha) in diluted spray	% of herbicide in Concentrate.					
	5	10	20	25	30	40
0.01	20	10	5	4	3.4	2.5
0.015	30	15	7.5	6	5.1	3.75
0.02	40	20	10	8	6.8	5.0
0.03	60	30	15	12	10.2	7.5
0.04	80	40	20	16	13.6	10.0
0.05	100	50	25	20	16.7	12.5
0.10	200	100	50	40	33.4	25
0.20	400	200	100	80	66.8	50
0.30	600	300	150	120	100.2	75
0.40	800	400	200	160	177.6	100
0.50	1000	500	250	200	167.0	125

ml or gram of active ingredient desired for 10 litres of water.

For Example : If one wants to use 0.20 kg a.i./ha through 5% herbicide concentrate, he needs 400 herbicide solution.

Appendix 11

Noxious Weeds Entered into India through Crop Seed and Packing Material

Weed species	Place of origin	Year of introduction into India
Argemone mexicana (Prickly pear)	Mexico	17th century
Lantana camera (wild sage)	Central America	1809
Eichhornia crassipes (Water hyacinth)	Brazil	1914
Euphorbia odoratum (Smeling Euphorbia)	Jamaica	1845
Acanthospermum hispidum	Northern Brazil or Central America	1946
Phalaris minor (Little Canary grass)	Mexico	1961

Appendix 12

Requirement of Commercial product of liquid herbicides at various active ingredients

Required active ingredient (kg/ha)	Presence of a.i. (%) in liquid herbicide											
	90	80	75	70	60	50	40	30	25	20	10	5
	Commercial liquid formulation to be sprayed (litres) in 1 ha. present of liquid formulation to be sprayed (L/ha)											
1	2	3	4	5	6	7	8	9	10	11	12	13
0.5	0.55	0.62	0.66	0.72	0.83	1.0	1.25	1.66	2.0	2.5	5.0	10.0
1.0	1.11	1.25	1.33	1.45	1.66	2.0	2.50	3.33	4.0	5.0	10.0	20.0
1.5	1.66	1.87	2.00	2.18	2.49	3.0	3.75	5.00	6.0	7.5	15.0	30.0
2.0	2.22	2.50	2.66	2.90	3.32	4.0	5.00	6.60	8.0	10.0	20.0	40.0
2.5	2.78	3.13	3.33	3.63	4.15	5.0	6.25	8.33	10.0	12.5	25.0	50.0
3.0	3.33	3.75	4.00	4.35	4.98	6.0	7.50	9.99	12.0	15.0	30.0	60.0
3.5	3.89	4.38	4.66	5.08	5.81	7.0	8.75	11.66	14.0	17.5	35.0	70.0
4.0	4.44	5.00	5.32	5.80	6.64	8.0	10.00	13.32	16.0	20.0	40.0	80.0
4.5	5.00	5.63	6.00	6.53	7.50	9.0	11.25	15.00	18.0	22.5	45.0	90.0
5.0	5.55	6.25	6.67	7.14	8.33	10.0	12.50	16.70	20.0	27.5	55.0	110.0
5.5	6.11	6.88	7.32	8.00	9.13	11.0	13.75	18.32	22.0	27.5	55.0	110.0
6.0	6.66	7.50	7.98	8.70	7.96	12.0	15.00	10.00	24.0	30.0	60.0	120.0

For example : If one wants to use 3.5 kg a.e. ha through 30 EC formulation, he needs 11.66 litres of commercial product.

Appendix 13

Seasons & its Characteristics

Month	*Approximate dates*	*Season*	*Characteristics of nature*
Chaitra Vaisakha	5th April—4th May 5th May—2nd June	Vasanth (Spring)	New leaves and buds are formed in Mango
Jyeshta Aashada	3rd June—1st July 2nd July—31st July	Greeshma (Summer)	High temperatures dries the plant life.
Shravana Bhadrapada	1st Aug.—29th Aug. 30th Aug.—27th Sept.	Varsha (Rainy)	Heavy down pour of rains
Aswiyuja Kartika	28th Sept.—27th Oct. 28th Oct.—25th Nov.	Sharad	Brilliant Moonlight
Margasira Purshya	26th Nov.—25th Dec. 26th Dec.—24th Jan.	Hemantha (Winter)	Dew fall during nights
Magha Phalguna	25th Jan.—23rd Feb. 24th Feb.—25th Mar.	Sasir (Autumn)	Fall of ripened leaves, mild cold nights

BIBLIOGRAPHY

Reference are mint tabs that flavour and freshen our thoughts

Abdul Waheb, A.S. and Rice E.L. 1967. Plant inhibition by Johnson grass and its possible significance in old field succession. *Bull. Torry Bot. Club.* 94-486.

Aldrich R.J. 1984. Weed-crop ecology. In : *Principles in Weed Management.* Breton (Massachusetts), 465 p.

Ampong-Nyarko, K. and De Datta, S.K. 1991. *A Hand Book for Weed Control in Rice.* IRRI, Manila, 113 p.

Anonymus 1998. The dropsy havoc. *The Hindustan Times.* Sept. 14, 1998, p. 5.

Akobundu, O. 1987. *Weed Science in the Tropics : Principles and Practices.* John Wiley and Sons, New York, 522 p.

Alstrom, S. 1990. *Fundamentals of Weed Management in Hot Climate Agriculture,* Uppasala 271 p.

Bell, D.T. and Kocppe, D.E. 1972. Noncompetative effects of gaint foxtail on the growth of corn. *Agron J.* **64** : 321-352

Bendall GM 1975. The allelopathic activity of california thistle (*Circium arvense* (L) Scop) in Tasmania. *Weed Res.* **15** : 77-81

Kumar, Binod 1991. M. Sc., Thesis. IARI, New Delhi.

Kumar Binod and Yaduraju N.T. 1992. *Plasticulture* **94** : 13-20.

Burkholder, P.R. 1952. Cooperation and conflict among primitive organisms. *Am Sci* **40** : 601-630.

Cardenas, J., Doli, J.,D. and Reyes C.E. 1992. *Argemone mexicana. Tropical Weeds* **1** : 243.

Charudattan, R. 1998. Plant pathogens for biological control of water hyacinth—A status report. In : Abstracts of papers, *First Global Working Group Meeting for the Biol. the integrated Control of Water Hyacinth* 10BC, *Harare, Zimbabwe*, p. 9.

Cassman, K.G., Harwood, R.R. 1995. The nature of agricultural systems: Food security and environmental balance. *Food Policy* **26** : 439-454.

Crosby, D.B. 1996. Impact of herbicide use on the environment. In : *Herbicides in Asian Rice*: *Transitions in Weed Management*, Naylor R. editor. Palo Alto, IRRI, Manila pp. 95-107.

Crosby, D.G. and Tutass, H.O. 1966. Photodecomposition of 2, 4-D. *J. Agric. Food Chem.* **14** : 596-99.

Dilday, R.H. 1990. Contribution of ancestral lines in the development of new cultivars of rice. *Crop Sci* **30**: 905-911.

Friedman, T. and Horowitz, M. 1971. Biologically active substances in subterranean parts of purple nutsedge, *weed Sci.* **19**: 398-401.

Fowler, M.C. and Robson, T.O. 1978. *Aquatic Bot.* **5**: 261-276.

Holm, L.G., Plucknett, D.L., Pancho, J.V. and Herberger, J.P. 1977. *The world's worst weeds, Distribution and biology.* An East-West Centre, The University Press of Hawaii, Honolulu, pp. *176-179.*

Hosmani M.M. and Habeeburrhman, P.V. 1992. *Soil Solarization a New Tool for Weed Control.* UAS, Dharwad pp. 15

Howrath, F.G. 1991. Environmental impacts of classical biological control. *Annu. Rev. Entomol* **36** : 485-509.

Heinrich, E.A., Medrano, F.G. 1984. *Leersia hexandra* a weed host of rice brown plant hopper (*Nilparvata lugens* (Stal). *Crop prot.* *3* : 77-85

Hess, F.D. 1992. Bioherbicides is an industry pesticide. *Plant prot. Qtrly.* *7* : 171.

Proceedings of symp. on Integrated weed Management for sustanable agriculture ***I***: 18-20 Nov, 1993, CCS Haryana Agril. Univ, 413 p.

Kropff, M.J. and Van laar, H.H. 1993. *Modelling crop. Weed interaction*, Wallingford (UK) : *CAB International*, 274 p.

Kanchan, S.D. and Chandra, J. 1979. Allelopathic effects of Parthenium hysterophorus, exudation of inhibitors through roots. *Plant and soil Sci.* **53** : 27-35.

Leather, G.R. 1983. Sunflowers (*Helianthus annus*) allelopathic to weeds. *Weed Sci* **31** : 37-42.

Mehta, I. and Sharma, R.K. 1972. *Hyacinth control J.* **10** : 16-19.

Moody K 1991. Crop-Weed complexity in lowland rice. In : *Proc of 13th Asian-Pacific Weed Sci Soc. Conf.*, Jakarta pp. 15-26.

Myrdal, G. 1968. *Asian Drama—An enquiry into the poverty of Nation* Vol. **II**, New York.

NAS 1975. *Under exploited tropical plants with promising economic value*. National Academy of Science, Washington, DC 172 p.

Pingali, P.L. Roger, P.A. 1995. *Impat of pesticides on farmer health and the rice environment*. Boston (Massachusett) Kluwer Academic Publishers, 664 p.

Raju, R.A. 1998. *Prevalent weed Flora in Peninsular India*. Allied Publishers, New Delhi 271 p.

Raju, R.A. 1996. *Ecology of Aquatic Weeds and Their Control*, Kalyani Publishers, Ludhiana 168 p.

Raju, R. A. 2000. *Ethnobotany of Rice Weed in South Asia*, Today and Tomorrow's Publishers, Delhi : 210 p.

Raju, R.A. 2000. *Glimpses of Rice Technology*. Agrobios (India) Jodhpur, 210 p.

Rola, A. and Pingali, P.L. 1993. *Pesticides, Rice Productivity and Farmer's Health. An Economic Assessment*, Manila : IRRI and Washington (DC) : World Resources Institute. 100 p.

Radosevich, S.R., Holt, J.S. 1984, *Weed Ecology—Implications for Weed Management*. Wiley and Sons, New York, 264 p.

Roger, P.A. and Bhviyan, S.I. 1990. Ricefield ecosystem management and its impact on disease vectors. *Water Resource Dev.* **6**(1) : 2-18.

Rice, E.L. 1974. *Alelopathy*. Academic Press. New York, 353 p.

Rao, V.S. 1983. *Principles of Weed Science*. Oxford and IBH Publishing House. New Delhi 540 p.

Sen, D.N. 1981. *Ecological Approaches to Indian Weeds*. Geobios International, Jodhpur 301 p.

Thakur, C. 1984 *Weed Science*. Metropolitan Publishers, New Delhi 495 p.

Timson, S.D. 1938. Humus and Witch Weed. *Rhod Agric J* **8** : 5-12.

Veki, K. 1983. Weed Succession by long-term application of herbicides in Japan. In : *IV PAC Pesticide chemistry. pp. 319-324.*

Whittakar, R.H., Feeny, P.P. 1971. *Allelochemics : Chemical Interactions Between Species, Science* **171** : 757-770.

World Bank, 1996. *Commodity Markets and the Developing Countries*. Washington (DC) 36P.

Worsham, A.D., Klingman, G.C. and Moreland D.E. 1962. Promotion of germination of *Striga asiatica* seeds by coumarin derivatives and effects on seedling development, *Nature* **195** : 199-201.

Yaduraju, N.T. 1975. *Studies on Striga Control in Sorghum*, M.Sc., Thesis, UAS, Bangalore.

Yaduraju, N.T. and Hosmani, M.M. 1979 *Striga asiatica* control in Sorghum, *PANS* **25** : 163.

Yoshikawa, F., Worsham A.D., Moreland, D.F. and Eplee, R.E. 1978. Biochemical requirements for seed germination and shoot development of witch weed. *Weed Sci.* **26** : 119-112.

Zumnno, N. 1977. Diseases of giant witch weed *Strica hermonthica* in West Africa. *Plant disease Rep.* **61** : 428-420.